TECHNIQUES IN MOLECULAR BIOLOGY

TECHNIQUES IN MOLECULAR BIOLOGY

Edited by JOHN M. WALKER and WIM GAASTRA

MACMILLAN PUBLISHING COMPANY
New York

Macmillan Publishing Company, 866 Third Avenue, New York,
New York 10022

Library of Congress Cataloging in Publication Data
Main entry under title:

Techniques in molecular biology.

Bibliogaphy: p.
Includes index.
1. Molecular biology ––Methodology. 2. Genetic
engineering––Methodology. I. Walker, John M.,
1948-   . II. Gaastra, Wim.
QH506.T4   1983   574.8'8'028   83–9419
ISBN 0–02–949830–9

Printed and bound in Great Britain

CONTENTS

PREFACE

The last few years have seen the rapid development of new methodology in the field of molecular biology. New techniques have been regularly introduced and the sensitivity of older techniques greatly improved upon. Developments in the field of genetic engineering in particular have contributed a wide range of new techniques. The purpose of this book therefore is to introduce the reader to a selection of the more advanced analytical and preparative techniques which the editors consider to be frequently used by research workers in the field of molecular biology. In choosing techniques for this book we have obviously had to be selective, and for the sake of brevity a knowledge of certain basic biochemical techniques and terminology has been assumed. However, since many areas of molecular biology are developing at a formidable rate and constantly generating new terminology, a glossary of terms has been included.

The techniques chosen for this book are essentially based on those used in a series of workshops on 'techniques in molecular biology' that have been held at The Hatfield Polytechnic in recent years. In choosing these chapters we have taken into account many useful suggestions and observations made by participants at these workshops. Each chapter aims to describe both the theory and relevant practical details for a given technique, and to identify both the potential and limitations of the technique. Each chapter is written by authors who regularly use the technique in their own laboratories. This book should prove useful to final year undergraduate (especially project) students, postgraduate research students and research scientists and technicians who wish to understand and use new techniques, but who do not have the necessary background for setting-up such techniques. Although lack of space precludes the description of in-depth practical detail (e.g. buffer compositions, etc.), such information is available in the references cited.

J.M. Walker
W. Gaastra

ABBREVIATIONS

A_{260}	—	Absorbance at 260nm
ABM	—	Aminobenzyloxymethyl
amp-r	—	ampicillin resistant
amp-s	—	ampicillin sensitive
APT	—	Aminophenylthioether
bis	—	methylenebisacrylamide
bp	—	base-pairs
CBA	—	Cyanogen bromide activated
CCC	—	Covalently closed circular (DNA)
cDNA	—	Complementary DNA
Dansyl	—	dimethylaminonapthalene-5-sulphonyl
DBM paper	—	diazobenzyloxymethyl paper
DEAE	—	diethylaminoethyl
DNA	—	deoxyribonucleic acid
DNase	—	deoxyribonuclease
DPT	—	diazophenylthioether
ds-cDNA	—	double-stranded cDNA
EDTA	—	ethylenediaminotetraacetic acid
HART	—	hybrid-arrest translation
HFBA	—	heptafluorobutyric acid
HPLC	—	High performance liquid chromatography
IEF	—	isoelectric focussing
mRNA	—	messenger RNA
nt	—	nucleotides
ODS	—	octadecylsilane
PAGE	—	polyacrylamide gel electrophoresis
PITC	—	phenylisothiocyanate
poly (A)	—	poly(adenylic)acid
poly A^+ RNA	—	polyadenylated RNA
poly (U)	—	poly(uridylic)acid
PTH	—	phenylthiohydantoin
RNA	—	ribonucleic acid
RNase	—	ribonuclease
RNP	—	ribonucleoprotein
RPC	—	reversed phase chromatography
rRNA	—	ribosomal RNA
S	—	Svedberg unit (corresponding to a sedimentation co-efficient of $1 - 10^{-13}$ seconds)
SDS	—	sodium dodecyl sulphate (sodium lauryl sulphate)
TCA	—	trichloroacetic acid

TEMED	—	N,N,N',N',-Tetramethylethylenediamine
tet-r	—	tetracycline resistant
TFA	—	trifluoracetic acid
Tris	—	Tris(hydroxymethyl)aminomethane
tRNA	—	transfer RNA
u.v.	—	ultra-violet

1 HIGH PERFORMANCE LIQUID CHROMATOGRAPHY OF PROTEINS AND NUCLEIC ACIDS

Keith Gooderham

CONTENTS

1. Introduction

High performance liquid chromatography (HPLC) provides an extremely effective method for separating complex mixtures of molecules. In addition to its high resolving power, HPLC also possesses at least three other major advantages over more traditional fractionation techniques: (i) separations are achieved very rapidly, usually in minutes rather than over a number of hours or days; (ii) only very small quantities of material are required for analysis, separations usually being obtained with nano- or picogram amounts rather than with micro- or milligram quantities; (iii) the fractionated molecules are usually detected by monitoring the u.v. absorbance of the column eluant and this passive detection method therefore allows fractions to be collected and the separated components recovered, on a preparative or at least semi-preparative scale.

Figure 1.1: A Schematic Representation of a High Performance Liquid Chromatography System

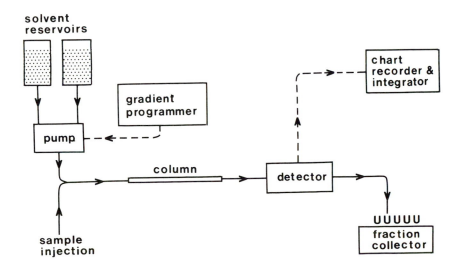

Note: Sample and solvent flow are shown by solid lines (———) and electronic signals by broken lines (— — — — —).

Although the basic layout of the HPLC system (Figure 1.1) is superficially similar to more traditional chromatography systems, there are a number of important differences. Extremely narrow bore tubing (1/64″ internal diameter) and fine column packings (5–20 micron particles) are used in order to obtain the rapid, high-resolution separations characteristic of HPLC. As a consequence of these modifications there is a considerable resistance to the flow of solvent through the system. In order to overcome this back-pressure, the solvent has to be pumped through the system under high pressure (up to 8,000 psi, though more usually between 500 and 2,000 psi). The use of high pressures has frequently resulted in HPLC being referred to as high pressure liquid chromatography.

The most important component of any HPLC system is the column, or more precisely the packing material it contains. A variety of different forms of chromatography are used in HPLC but all of these methods depend upon the relative importance of the competing attractions of the packing material, or *stationary phase*, and the eluting solvent, or *mobile phase*, for the sample. These interactions can then be modified in a number of ways, for example by modifying the composition of the mobile phase, as in gradient chromatography, or by changing the flow rate (pressure) or temperature at which the chromatography is performed.

Currently three major types of HPLC are commonly employed. These are: size-exclusion chromatography; reversed-phase chromatography and ion-exchange chromatography. Ideally each of these methods will separate the sample on the basis of only one property, i.e. size, hydrophobicity and charge, respectively. However, ideal systems are rarely encountered in experimental work and any given separation is unlikely to be due to only one of these parameters. This situation is further complicated by the multiplicity of manufacturers producing HPLC columns. Differences in particle size and shape, as well as the choice of capping groups (see Section 3), can all have a profound effect upon the resulting chromatography.[1] Consequently, when attempting to follow a previously published method, it is important not only to reproduce the precise operating conditions but also to use an identical column.

Though commercially prepared columns are now relatively inexpensive, it is also possible to prepare satisfactory columns in the laboratory, using either commercially available packing machines or by using a home-made apparatus (for example as described by Manius and Tscherne[2]). In addition a number of companies now offer a highly competitive column packing service for a wide range of stationary phases. Once a suitable column has been obtained its working life can be considerably extended by filtering the mobile phase and sample prior to use, or by using a guard column positioned directly in front of the main column. Both of these precautions are designed to filter out any particulate matter, which otherwise would eventually block the column.

2. Size-exclusion Chromatography

Size-exclusion chromatography (also known as steric-exclusion, gel filtration and gel permeation chromatography), has long been an important method for the separation of proteins and peptides. However, it is only comparatively recently that this method has been successfully applied to HPLC. The principal reason for this delay has been the difficulty of obtaining a packing material which is both porous and yet of sufficient mechanical strength to withstand the pressures normally associated with HPLC. In addition, the majority of these column packings have a silica base, which is very reactive owing to the presence of negatively charged silanol groups. The stationary phase will therefore act as a weak cation-exchanger interfering with the size-dependent separation by binding positively charged molecules.

Advances in column technology have, however, largely overcome these problems. Modern columns are now capable of operating at relatively high pressures (up to 2000 psi or greater) and at a flow rate of 1 to 2 ml/minute. Ionic interactions between the stationary phase and the sample have also been reduced by the use of small organic 'capping' groups (see Section 3), though in some instances this has only replaced the ionic component of the chromatography by a reversed-phase element. The resulting chromatography can be further improved by a judicious choice of ionic strength and pH when selecting the mobile phase. In general, increasing the ionic strength of the buffer will minimise the degree of interaction between the sample and any remaining silanol groups. The pH of the mobile phase will similarly influence the binding of the sample to the charged stationary phase by controlling the degree of ionization and net charge of the sample.

Although ion-exchange and reversed-phase effects can both influence the separation, the resulting fractionation primarily depends on the molecular weight and shape of the sample molecules. Large molecules are excluded from entering the porous stationary phase and are carried straight through the column, while progressively smaller molecules are increasingly able to enter into the stationary phase and consequently have proportionately longer elution times. The porosity of the stationary phase therefore determines the molecular weight range which may be analysed and, as with the more traditional gel filtration media, a number of different fractionation ranges are required to meet the majority of possible applications. A list of the major size-exclusion columns suitable for protein and nucleic acid research, together with their theoretical fractionation ranges, is presented in Table 1.1.

While the majority of these columns are intended for use at high flow rates (0.5 to 2.0 ml/min) certain manufacturers (for example, LKB Instruments Ltd) are currently producing a series of columns which are intended to be operated at extremely low flow rates (~50 μl/min). The rationale behind this approach is that at low flow rates the individual components of the sample have a greater opportunity to interact with the column and the resulting separation, though taking several hours, is therefore of considerably higher resolution.[3] Most workers at the moment, however, use the high flow rate columns and it remains to be seen whether the increased resolution obtained with these columns will lead to their wider use.

Table 1.1:   Fractionation Ranges of some of the Major High Performance Size-exclusion Chromatography Columns

	Fractionation range (daltons)	
	Native globular proteins	Denatured random coil proteins
Group A[a]		
I–60	1,000–20,000	600–8,000
I–125	2,000–80,000	1,000–30,000
I–250	10,000–500,000	2,000–150,000
Group B[b]		
TSK G2000SW[c]	500–60,000	1,000–25,000
TSK G3000SW	1,000–300,000	2,000–70,000
TSK G4000SW	5,000–1,000,000	3,000–400,000

Notes:

a. Columns in group A are manufactured and supplied by Waters Associates Inc. (Milford, Mass. 01757, USA).

b. Group B columns are manufactured and supplied by Toyo Soda Manufacturing Co. Ltd (Tokyo, Japan).

c. LKB Instruments Ltd and Bio Rad Laboratories also use these TSK column packings for their Ultropac and Bio Sil columns respectively.

The principal application of size-exclusion HPLC, has been for the separation of proteins, using both non-denaturing and denaturing aqueous solvents. For non-denaturing chromatography, phosphate buffers ($\sim$ pH7), and usually containing 0.1–0.3M NaCl to minimise secondary ionic interactions, have been widely used (see Figure 1.2 and examples given in references[4,5,6,7]). Although satisfactory separations have been obtained under these conditions, the denaturing solvent systems generally produce superior separations and are preferred, providing it is not important to maintain the proteins in their native state. Denaturing solvent systems have the considerable advantage of permitting the analysis of a wider range of proteins by overcoming many potential problems caused by insolubility and aggregation of the proteins. Molecular weight determinations are also, in general, more accurate under these conditions because the denatured proteins assume an open random coil structure. In contrast, native proteins have a complex globular conformation and this frequently causes an anomalous rate of migration through the column. Two principal denaturants have been used: guanidine hydrochloride[8,9] and the ionic detergent, sodium dodecyl sulphate. The latter has been particularly effective in improving the resolution of mixtures of proteins.[10,11]

Though the major application of size-exclusion chromatography has been in the fractionation of proteins, it seems likely that this technique will be of value in the analysis of nucleic acids, including ribosmal and transfer RNA's[12] and short synthetic oligonucleotides.[13]

Figure 1.2: Non-Denaturing High Performance Size-exclusion Chromatography of Protein Standards

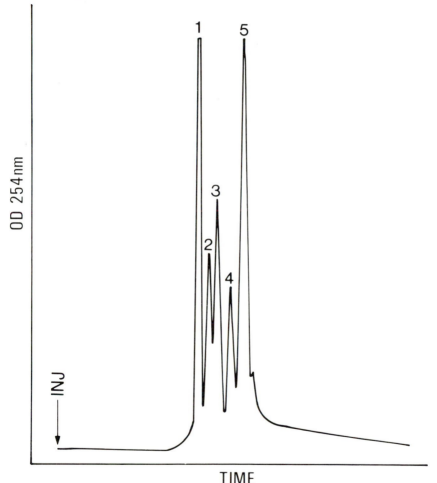

Note: 50 μl of a standard mixture of 5 proteins, containing 10 μg of each protein (1. ferritin, molecular weight (MW) 540,000, retention time (RT) 9.47'; 2. phosphorylase b, MW 94,000, RT 10.34'; 3. albumin, MW 67,000, RT 10.89'; 4. trypsin inhibitor, MW 20,100, RT 11.97'; 5. carbonic anhydrase, MW 30,000, RT, 12.91') was separated on two Waters I125 columns (coupled in series) with 50 mM sodium phosphate (pH 6.8) at a flow rate of 1 ml/min and the absorbance was monitored at 254nm.

3. Reversed-phase Chromatography

Reversed-phase chromatography (RPC) is probably the most versatile and widely used of all current forms of HPLC. The precise separation mechanism is as yet unclear, but it is likely that it is a form of partition chromatography, based upon the relative affinity of the sample for the saturated hydrocarbon molecules of the stationary phase and the competing attraction of the polar mobile phase. A variety of saturated hydrocarbons of differing chain lengths are used as the stationary phase, ranging from short chain alkyl groups to larger molecules containing as many as 18 carbon atoms, such as the widely used octadecyl silane (ODS or C_{18}) columns. The short chain alkyl and intermediate C_8 columns are preferred where more polar samples are to be analysed but there is a considerable overlap between these columns and the ODS columns.

The majority of these stationary phases are covalently bound to a microparticulate silica glass support. Ideally this support should not in itself contribute to the chromatography; however, the initial derivatisation of the silica is seldom complete and reactive silanol groups frequently remain. These groups not only interfere with the chromatography but will also cause the hydrolysis and eventual destruction of the silica support. Most manufacturers now attempt to overcome this problem by a secondary treatment which is designed to cap these reactive sites with small hydrocarbon molecules. The efficiency of the initial derivatisation, together with the subsequent choice of capping group (if any), is to a large degree responsible for the differences between columns from different manufacturers. Where unreacted silanol groups still remain, 50 mM tetramethylammonium hydroxide may be added to the mobile phase. The resulting tetra methyl ammonium cation will then block any residual silanol groups.[14]

The mobile phase is usually a combination of a weak aqueous buffer or dilute acid and a water miscible organic solvent. A variety of organic solvents are used in RPC, the most frequently encountered being methanol and where a more strongly non-polar organic phase is required, acetonitrile. The chromatography can either be isocratic (i.e. the composition of the mobile phase remains constant throughout the run), or the percentage of the organic phase can be increased during the analysis, progressively eluting more and more tightly bound fractions from the stationary phase. Ideally, isocratic separations are favoured whenever possible as they do not require complex gradient systems, are easily reproduced and eliminate the necessity for re-equilibrating the column between runs. Where a satisfactory separation cannot be obtained by this method a gradient system must then be used. A simple linear gradient is preferable, though complex samples frequently demand the use of a more complicated gradient. Irrespective of whether an isocratic or gradient elution is performed, care should be taken to ensure that the organic phase does not cause the precipitation of any of the aqueous phase components. If this does occur, the resulting precipitate can block the column as well as the associated tubing and detector flow cell, all of which are very difficult to clear.

Although the organic component of the mobile phase is of primary importance in eluting the sample from the stationary phase, the aqueous component is largely responsible for the initial binding of the sample. In particular the pH of the mobile phase can be used to control the ionisation of the sample. At a pH where the ionisation of the sample is suppressed, the sample will have a proportionately greater non-polar character and will therefore bind more strongly to the stationary phase, while at a pH where the ionisation of the sample is favoured the resulting interaction with the hydrophobic stationary phase is reduced.

Another important variable in RPC is the choice of the temperature at which the chromatography is performed. In general, an increase in temperature will result in a more rapid elution of the sample as well as resulting in an overall sharpening of the individual peaks. The capacity of the column also increases at higher temperatures and this can be exploited for preparative applications, where a larger initial sample is required than in routine analytical work.

The considerable versatility and power of RPC has resulted in this technique becoming the preferred method for tackling many problems in both protein and nucleic acid biochemistry. RPC has proved to be a particularly valuable tool for both analytical and preparative fractionations of complex mixtures of peptides. Initially RPC was limited to the separation of peptides rich in hydrophobic residues, particularly polypeptide hormones.[15,16] This relatively restricted use of RPC has now been extended to a far wider range of applications. Of particular significance is the contribution RPC has made to both peptide mapping and the isolation of peptides for protein sequencing studies (see also Chapter 4). Traditionally paper or thin-layer chromatography and high-voltage electrophoresis have been the standard methods used for producing peptide maps. These methods are, however, not only time-consuming but they are also relatively insensitive and the separated peptides are not readily carried forward for further analysis.

In marked contrast, by using RPC, high resolution separations are achieved rapidly (in less than two hours).[17,18] In addition, a wide variety of methods may be used to detect the eluting peptides, including u.v. absorbance,[19] fluorimetry[20,21,22] and scintillation counting.[23] More significantly, the eluting peptides may be recovered and immediately subjected to further analysis. It is this preparative capacity which makes RPC such a potentially valuable tool for the isolation of peptides for protein sequencing studies. A number of workers have already successfully used this technique[24,25] and with the increasing trend towards micro-sequencing, HPLC and RPC in particular are likely to assume an even greater importance.

Another area of protein biochemistry where RPC has already made a significant impact is in the identification of phenylthiohydantoin (PTH) amino acids obtained from the automatic protein sequenator (see Chapter 5). In contrast to other methods RPC is not only very simple, quantitative and sensitive but is also extremely rapid (average analysis time 20 minutes). These advantages combine to make RPC the preferred method for PTH amino acid analysis. By using RPC continuous evaluation of the sequenator run is possible, ensuring that the maximum amount of information is obtained while minimising the use of valuable machine time and chemicals. Also, because the PTH amino acids are analysed rapidly and directly, the relatively unstable PTH derivatives of tryptophan, glutamine, asparagine, serine and threonine can all be identified more easily.

A large number of different methods have been published for the identification of PTH amino acids by RPC. The majority of these methods use an ODS column equilibrated against acetate buffer (pH3.7–5.4). The PTH amino acids are then eluted with an acetonitrile or methanol gradient[26,27,28] or under isocratic conditions.[29] Excellent separations have also been obtained using cyanopropylsilane (CN) reversed-phase columns.[30] The CN columns are reported to be superior to comparable ODS columns in that the reproducibility and sensitivity of the separations is significantly improved and the columns have a considerably longer working life (2,500–3,000 injections for a CN column as against 750–1,500 injections for an ODS column).[31]

The separation of as many as 22 PTH amino acids in approximately 20 minutes clearly shows the considerable potential of HPLC and RPC in particular. As a result of these successes, efforts are now being made to apply this technique to the analysis of a number of other amino acid derivatives (dansyl amino acids[32] and dimethylaminoazobenzene-thiohydantoins[33] obtained from manual peptide sequencing studies (see Chapter 5) as well as to the underivatised amino acids.[34]

Apart from its use in protein biochemistry RPC is also an important technique for both the analytical and preparative fractionation of nucleic acids. In particular, RPC offers an alternative to ion-exchange chromatography (see Section 4) for the quantitative analysis of nucleotides, nucleosides and bases from both DNA and RNA. Excellent high resolution separations can be obtained for all of these groups, including their methylated derivatives.[35,36] Synthetic and natural oligo-nucleotides may also be separated by RPC.[37] The stationary phase RPC-5 has been particularly valuable in this respect and has been widely used for the analysis and isolation of plasmids, supercoiled viral DNA's and DNA restriction fragments.[38,39,40]

Although RPC-5 is no longer available from the original manufacturer (Miles Laboratory, Elkhart, Indiana, USA), an acceptable alternative, RPC-5 Analog, is now available from Bethesda Research Laboratories Inc. Both RPC-5 and RPC-5 Analog rely on a combination of reversed-phase

Figure 1.3: High Performance Ion-exchange Chromatography of Nucleic Acid Bases

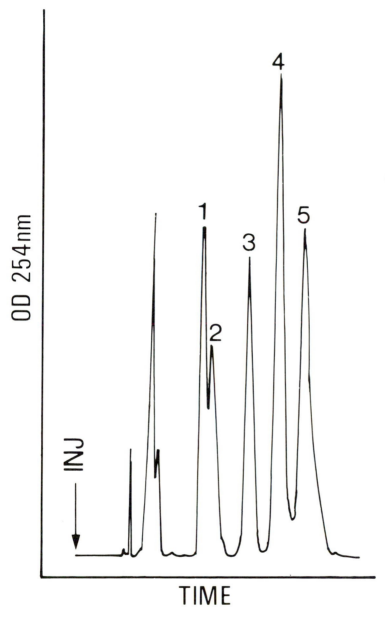

Note: 50 μl of a standard mixture of 5 bases, containing 0.5 pM (10^{-9} M) of each base (1. cytosine, retention time (RT) 6.30'; 2. 5-methyl cytosine, RT 6.81'; 3. thymine, RT 8.74'; 4. guanine, RT 10.26'; 5. adenine, RT 11.62') was separated on a 25 cm Zorbax SAX anion exchange column (Dupont) with 30 mM sodium phosphate (pH 4.5) at a flow rate of 1 ml/min and the absorbance was monitored at 254 nm.

and ion-exchange chromatography and the fractions are eluted by a simple linear salt gradient.

4. Ion-exchange Chromatography

In the previous two sections I have briefly discussed how interactions between charged groups in the stationary phase and the sample can modify the results expected when using these methods. These interactions can, however, be intentionally exploited in ion-exchange chromatography and they provide an alternative method to size-exclusion and reversed-phase chromatography. The stationary phase of an ion-exchange column may either be a positively charged cation-exchanger or a negatively charged anion-exchanger. These charged groups are neutralised by oppositely charged, counter-ions in the mobile phase, the counter-ions being replaced during the chromatography by more highly charged sample molecules. For some applications, the competition between the counter-ion and the sample for the charged groups of the stationary phase is sufficient to obtain a separation of the sample (for example see Figure 1.3). More frequently the exchange can only be reversed, and the fractionation of the sample achieved, by increasing the ionic strength and/or changing the pH of the mobile phase.

The principal application of ion-exchange chromatography has been in the analysis of nucleic acids and their constituent bases, nucleosides and nucleotides (the comprehensive, though dated review by Singhal[41] provides an invaluable analysis of the factors controlling their separation). The increasing use of RPC has, however, led to a progressive decline in the popularity of ion-exchange chromatography although excellent separations can be obtained using this technique.[42,43,44]

5. Conclusion

HPLC provides an extremely versatile method for the analysis of complex biological samples. Its ability to perform both analytical and preparative analyses rapidly, together with the variety of fractionation techniques, ensures that this method will assume an even greater importance in the future. In addition to the applications outlined in the previous sections, new applications will, without doubt continue to be developed (for example the use of HPLC in routine amino acid analysis).

Given the rate of progress in this field any review of the subject will inevitably rapidly become outdated and the reader is referred to the latest technical publications and specialist journals for a true picture of the state of the art.

References

1. A. Goldberg, 'Comparison of reversed phase packings', *Liquid Chromatography Technical Report*, Du Pont Company, Analytical Instrument Division.
2. G.J. Manius and R.J. Tscherne, 'Procedure for packing efficient octadecylsilane columns', *International Laboratory* (March 1981), 38–48.
3. J. Sjodahl, 'High-resolution liquid chromatography of proteins', *Science Tools*, 27 (1980), 54–56.
4. K. Fukano, K. Komiya, H. Sasaki and T. Hashimoto, 'Evaluation of new supports for high-pressure aqueous gel permeation chromatography: TSK-gel SW type columns', *J. Chromatog.*, 166 (1978), 47–54.
5. S. Rokushika, T. Ohkawa and H. Hatano, 'High-speed aqueous gel permeation chromatography of proteins', *J. Chromatog.*, 176 (1979), 456–461.
6. Y. Kato, K. Komiya, H. Sasaki and T. Hashimoto, 'Separation range and separation efficiency in high-speed gel filtration on TSK-gel SW columns', *J. Chromatog.*, 190 (1980), 297–303.
7. M.E. Himmel, P.G. Squire, 'High pressure gel permeation chromatography of native proteins on TSK-SW columns', *Int. J. Peptide Proteins Res.*, 17 (1981), 365–373.
8. R.C. Montelaro, M. West and C.J. Issel, 'High-performance gel permeation chromatography of proteins in denaturing solvents and its application to the analysis of enveloped virus polypeptides', *Anal. Biochem.*, 114 (1981), 398–406.
9. Y. Kato, K. Komiya, H. Sasaki and T. Hashimoto, 'High-speed gel filtration of proteins in 6M guanidine hydrochloride on TSK-gel SW columns', *J. Chromatog.*, 193 (1980), 458–463.
10. Montelaro, West and Issel, 'High-performance gel permeation chromatography'.
11. Y. Kato, K. Komiya, H. Sasaki and T. Hashimoto, 'High-speed gel filtration of proteins in sodium dodecyl sulphate aqueous solution on TSK-gel SW type', *J. Chromatog.*, 193 (1980), 29–36.
12. LKB Instruments Ltd. Application Note No. 2135:000 AP3 (1980).
13. D. Molko, R. Derbyshire, A. Guy, A. Roget, R. Teoule and A. Boucherle, 'Exclusion column for high-performance liquid chromatography of oligo-nucleotides', *J. Chromatog.*, 206 (1981), 493–500.
14. M.E.F. Biemond, W.A. Sipman and J. Olivie, 'Quantitative determination of polypeptides by gradient elution high pressure liquid chromatography', *J. Liquid Chromatogr.*, 2 (1979), 1407–1435.
15. K.A. Gruber, S. Stein, L. Brink, A. Radhakrishnan and S. Udenfriend, 'Fluormetric assay of vasopressin and oxytocin: A general approach to the assay of peptides in tissues', *Proc. Natl. Acad. Sci. USA*, 73 (1976), 1314–1318.
16. M.J. O'Hare and E.C. Nice, 'Hydrophobic high-performance liquid chromatography of hormonal polypeptides and proteins on alkylsilane bonded silica', *J. Chromatog. 171* (1979), 209–226.
17. J.M. Cecka, M. McMillan, D.B. Murphy, H.O. McDevitt and L. Hood, 'Partial N-terminal amino acid sequence analyses and comparative tryptic peptide maps of murine Ia molecules encoded by the I-A subregion', *Eur. J. Immunol.*, 9 (1979), 955–963.
18. J.B. Wilson, H. Lam, P. Pravatmuang and T.H.J. Huisman, 'Separation of tryptic peptides of normal and abnormal α, β, γ and δ hemoglobin chains by high-performance liquid chromatography', *J. Chromatog.*, 179 (1979), 271–290.
19. Ibid.
20. Gruber, Stein, Brink, Radhakrishnan and Udenfriend, 'Fluormetric assay of vasopressin and oxytocin'.
21. O'Hare and Nice, 'Hydrophobic high-performance liquid chromatography of hormonal polypeptides and proteins'.
22. J.L. Meek, 'Prediction of peptide retention times in high pressure liquid chromatography on the basis of amino acid composition', *Proc. Natl. Acad. Sci. USA*, 77, 1632–1636.
23. Cecka, McMillan, Murphy, McDevitt and Hood, 'Partial N-terminal amino acid sequence analyses'.
24. G.E. Gerber, R.J. Anderegg, W.C. Herlihy, C.P. Gray, K. Biemann and H.G. Khorana, 'Partial primary structure of bacteriorhodopsin: Sequencing methods for membrane proteins', *Proc. Natl. Acad. Sci. USA*, 76 (1979), 227–231.

25. G.J. Hughes, K.H. Winterhalter and K.J. Wilson, 'Micro-sequence analysis: 1. Peptide isolation using high-performance liquid chromatography', *FEBS Letts.*, *108* (1979), 81–86.

26. A.S. Bhown, J.E. Mole and J.C. Bennet, 'An improved procedure of high-sensitivity microsequencing: Use of aminoethylaminopropyl glass beads in the Beckman sequencer and the Ultra-sphere ODS column for PTH amino acid identification', *Anal. Biochem.*, *110* (1981), 355–359.

27. P.W. Moser and E.E. Rickli, 'Identification of amino acid phenylthiohydantoins by gradient high-performance liquid chromatography on Spherisorb S5-ODS', *J. Chromatog.*, *176* (1979), 451–455.

28. C.L. Zimmerman, E. Appella and J.J. Pisano, 'Rapid analysis of amino acid phenylthiohydantoins by high-performance liquid chromatography', *Anal. Biochem.*, *77* (1977), 569–573.

29. G.E. Tarr, 'Rapid separation of amino acid phenylthiohydantoins by isocratic high-performance liquid chromatography', *Anal. Biochem.*, *111* (1981), 27–32.

30. N.D. Johnson, M.W. Hunkapiller and L.E. Hood, 'Analysis of phenylthiohydantoin amino acids by high-performance liquid chromatography on DuPont Zorbax Cyanopropylsilane columns', *Anal. Biochem.*, *100* (1979), 335–338.

31. Ibid.

32. S. Weiner and A. Tishbee, 'Separation of DNS-amino acids using reversed-phase high-performance liquid chromatography a sensitive method for determining N-termini of peptides and proteins', *J. Chromatog.*, *213* (1981), 501–506.

33. J.Y. Chang, A. Lehman and B. Wittmann-Liebold, 'Analysis of dimethylaminobenzene-thiohydantoins of amino acid by high-pressure liquid chromatography', *Anal. Biochem.*, *102* (1980), 380–383.

34. R. Schuster, 'Determination of free amino acids by high-performance liquid chromatography', *Anal. Chem.*, *52* (1980), 617–620.

35. C.E. Salas and O.Z. Sellinger 'Rapid quantitative separation by high-performance liquid chromatography of methylated bases in transfer RNA', *J. Chromatog.*, *133* (1977), 231–236.

36. I. Wagner and I. Capesius; 'Determination of 5-methyl-cytosine from plant DNA by high-performance liquid chromatography', *Biochem. Biophys. Acta*, *654* (1981), 52–56.

37. G.D. McFarland and P.N. Borer, 'Separation of oligo-RNA by reverse-phase HPLC', *Nucleic Acids Res. 7* (1979), 1067–1080.

38. A.N. Best, D.P. Allison and G.D. Novelli, 'Purification of supercoiled DNA of plasmid Col El by RPC-5 chromatography', *Anal. Biochem. 114* (1981), 235–243.

39. R.K. Patient et al. 'Influence of A-T content on the fractionation of DNA restriction fragments by RPC-5 column chromatography', *J. Biol. Chem. 254* (1979), 5548–5554.

40. R.D. Wells et al. 'RPC-5 column chromatography for the isolation of DNA fragments', *Methods in Enzymology, 65* (1980), 327–347.

41. R.P. Singhal, 'Separation and analysis of nucleic acids and their constituents by ion-exclusion and ion-exchange column chromatography', *Separation and Purification Methods, 3* (1974), 339–398.

42. Du Pont Company, Hertfordshire, U.K., Analytical Instruments Division. LC Column Report. Methods Development Guide.

43. Whatman Inc. 'Analysis of nucleic acid constituents by high-performance liquid chromatography', *Bulletin No. 116* (1976).

44. H. Yuji, H. Kawasaki, T. Kobayashi and A. Yamaji, 'Determination of cytosine, 3-methylcytosine, and 5-methylcytosine in nucleic acids by high performance liquid chromatography', *Chem. Pharm. Bull. 25* (1977), 2827–2830.

2 ANALYTICAL ELECTROPHORETIC TECHNIQUES IN PROTEIN CHEMISTRY

Bryan John Smith and Robert H. Nicolas

CONTENTS

1. Introduction

The method often chosen today for analysis of proteins is electrophoresis in polyacrylamide gels. Microgram amounts of material can be used to give information on a protein's molecular weight, isoelectric point, etc. Many electrophoretic techniques have been described in the literature, but this chapter describes just some of the most commonly used of them. The general principles and practice of electrophoresis are also described but relate mainly to the methods discussed. For more comprehensive treatments of the subject, the reader is directed to reviews such as those by Smith[1] and Gordon[2] and references therein. Other references are given below for particular points not covered at length in these reviews.

2. Basic Principles of Electrophoresis in Polyacrylamide Gels

Many of the amino acid side chains of a protein in solution are capable of undergoing ionisation, so becoming positively or negatively charged. By this process a protein molecule may become cationic or anionic, according to the sum of the various charges on it. If such a charged particle is placed in an electric field (between two electrodes) it will move towards the electrode of opposite charge. This, put simply, is the process of electrophoresis. A detailed description of electrophoresis is given by Tanford[3] but it may be stated briefly that the following factors affect the migration (or electrophoretic mobility) of proteins undergoing electrophoresis:

(i) The charge on the protein – which dictates the direction of the migration and directly influences the velocity. Proteins are weak electrolytes and their ionisation is greatly affected by the pH of the surrounding medium. The charge on a protein in solution is controlled by the use of various buffer systems. Gordon[4] has discussed the use of buffers in electrophoresis at greater length.

(ii) Strength of the electric field – which directly affects mobility.

(iii) Frictional forces – which oppose migration. For example, an increase in viscosity of the medium through which electrophoresis is occurring decreases a protein's mobility. Again, a large spherical molecule will encounter more resistance to movement than will a small spherical one, as will a protein of extended conformation rather than a spherical one.

Some of these factors relate directly to properties of the proteins

27

themselves, so that under any one set of conditions of electric field strength, pH etc., two proteins of similar size and shape but of different amino acid composition (and therefore, charge) will have different electrophoretic mobilities. The process of electrophoresis, then, has the potential to separate such proteins from one another, for the purposes of analysis or purification.

Satisfactory resolution of mixtures of proteins is difficult to achieve if the system is in a totally mobile, liquid phase. Problems such as those due to diffusion and convection are reduced, however, by employing a stabilising or supporting medium, which immobilises part of the system. The relative merits of the various media available will not be discussed here,[5,6] but it may be stated that polyacrylamide is generally the medium of choice. In structure it is a three-dimensional matrix which is permeated by buffer to form a gel. Proteins can pass through the holes or pores in this matrix. One feature of a polyacrylamide gel which makes it an attractive choice of medium is the ease with which a wide variety of sizes of pore can be made. This has an important practical consequence because the pores in the gel act like a sieve, retarding the motion of larger molecules. A particular pore size may be well-suited to resolve a certain mixture of proteins by virtue of this 'molecular sieving' effect and so the ability to manipulate pore size is an important aid in improving the results of electrophoresis. It should be remembered, however, that molecular sieving is not the only factor affecting electrophoretic mobility in polyacrylamide gels (see above). For instance, a larger but more highly charged molecule may migrate faster in a gel of large pore size than does a smaller molecule of smaller charge, because of the overriding effect of charge. In a gel of smaller pore size their relative mobilities may be reversed by the increased effects of molecular sieving.

3. Construction of the Gel

The three-dimensional matrix of the polyacrylamide gel is made of polymers of acrylamide which are crosslinked together. The most common agent used for crosslinking is N,N'methylenebisacrylamide (or 'bis'). The acrylamide content of a gel (sometimes known as % T) is expressed as % w/v, that is, the total number of grams of acrylamide and 'bis' per 100 cm^3 of gel mixture before polymerisation. The amount of crosslinking agent used in a gel (% C) is given as the ratio of the crosslinking agent to the acrylamide monomer, expressed as % w/w.

Polymerisation of the acrylamide and 'bis' is a free radical chain reaction. The process is initiated by the generation of radicals, commonly by a mixture of ammonium persulphate (at about 0.3% w/v or less) and N,N,N',N'-tetramethylethylenediamine (TEMED, at about 0.2% v/v or less). The persulphate activates the TEMED, leaving it with a reactive

Figure 2.1: Formation of Polyacrylamide

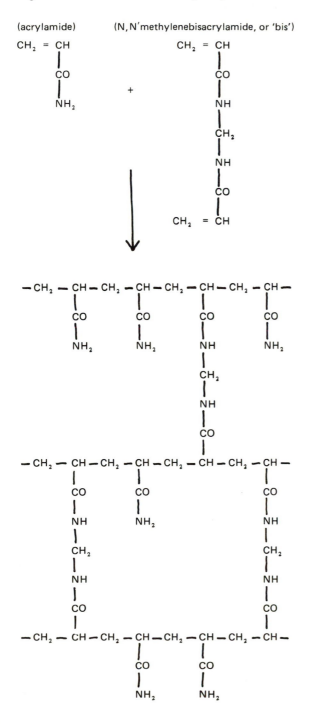

unpaired electron. This radical then reacts with an acrylamide molecule to produce a new radical which can react with *another* acrylamide molecule, and so on to build up a polymer. Occasionally one end of the symmetrical 'bis' molecule is incorporated in the polymer, and if the other end of that 'bis' molecule is included in another polymer molecule, then a crosslink has been formed (see Figure 2.1).

The formation of crosslinks occurs at random during polymerisation and the result is a complex tangle of polymer chains linked to each other at various points. Spaces of nm proportions interconnect within this tangle to allow movement of electrolytes through the gel during electrophoresis. The actual sizes of the pores vary throughout the gel, but the average size varies according to the total amount of acrylamide in the gel and also according to the ratio of acrylamide to 'bis'. Thus, the average pore size will be smaller in a 20% polyacrylamide gel than in a 15% gel containing the same % 'bis'. The ratio of 'bis' to acrylamide has an optimum of 5% w/w for achieving small average pore size. Either more or less 'bis' than 5% results in larger pores, whatever the total amount of acrylamide present. The polyacrylamide content of a gel may be anything from 30% to 3%. The very fragile 3% T gels may be strengthened by addition of agarose (say, 0.5% w/v). The lower % polyacrylamide gels are useful for separation of large proteins (of molecular weight of the order of 10^6). Small proteins are better resolved in higher percentage polyacrylamide gels but the minimum size is about 5,000 daltons and peptides smaller than this are perhaps better dealt with by electrophoresis in thin layer or paper media.

Polyacrylamide gels are usually cast in one of two basic shapes, namely the tube (narrow cylinder) and the thin slab. The most important difference between them is the number of samples which can be run on each system. On a tube gel this number is only one or two, but the number on a slab is dictated largely by the size of the apparatus – 25 samples on a 12 cm-wide slab is possible. The slab can be more economical and better for comparison of different samples under identical conditions of electrophoresis, and consequently is a popular choice. What follows below applies particularly to the slab gel type but also in principle to the tube gel. The thickness of the gel is very variable but is limited by the need to keep the gel cool (to minimise diffusion of the protein zone). Gels of more than a few mm thickness may require special provision for cooling (see below).

Ready-made gels are available commercially but are easy to make in the laboratory. Both acrylamide and 'bis' are fairly stable solids at room temperature in the dark (in brown bottles) although after several years spontaneous polymerisation and hydrolysis may prove a problem. Aqueous solutions of these agents, either together or separately, are stable for several months in the cold (4°C) and dark. A stock solution of ammonium persulphate (say, at 10% w/v) is stable for a few weeks in the cold and dark. Buffer solutions, too, may be stable enough to allow the use of stock solutions to make gel preparation more convenient. It is advisable to filter solutions before use, for the presence of particles of dust or other insoluble matter in a gel can distort the zone or band of protein as it migrates past them.

The gel is cast in the container in which it is used, so that it fits tightly enough to prevent leakage of the proteins around the side of the gel. Before use this container is thoroughly cleaned with detergent, solvents and water to remove contaminants. It is usually made of glass, which allows observation of the gel. Slab gels are cast between two glass plates which are separated by spacers of plastic or other material (see Figure 2.2). These spacers can be held in place with white petroleum jelly. For vertical electrophoresis the plates are sealed across the bottom (with tape or plastic strip) to contain the liquid while it polymerises. This seal is then removed to expose the gel for use. For horizontal electrophoresis in slab form, one of the glass plates may be removed, to expose one side of the gel.

Figure 2.2: An Apparatus for Running Slab Polyacrylamide Gels

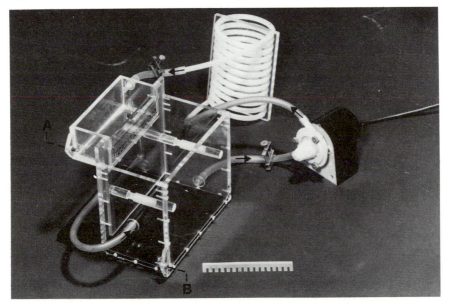

Note: The design is principally as described by Studier[7] but with provision for a deeper lower reservoir. The pump circulates the buffers in the direction indicated by arrows. The coil (when immersed in water or ice) allows efficient cooling of circulating buffers. The gel is held in place by two screw clamps. Wells in the gel are formed by the template shown in the foreground. Electrodes extend across the apparatus from terminals A and B.

To cast a gel the various constituents (except the persulphate and TEMED) are mixed to give the concentrations of acrylamide and buffer appropriate to the type of gel. The solution is then degassed to remove oxygen and to discourage formation of air bubbles when the solution is poured. This is because oxygen can mop up radicals, and so inhibit polymerisation. Even a small air bubble in the solution will inhibit polymerisation around it thus generating a large discontinuity in the gel which may distort a protein band. When the solution has been degassed the persulphate and TEMED are added, with mixing, to initiate the polymerisation. The solution is then gently poured between the glass plates and immediately overlayered with a few mm-deep layer of distilled water, or better, isobutanol, which is less dense than water. This procedure excludes air and also gives a very flat surface to the gel when it has polymerised, at which time the layer of isobutanol or water is washed off the gel. It may be necessary to make provision for loading samples onto the gel. This is done by making wells or slots in one end of the gel. These wells are usually as thick as the gel, are 3 mm or so wide and of variable depth, and are separated from each other by about 2 mm or more. To make them, a template (see Figure 2.2) is immersed in the gel solution before it is polymerised. This by itself may exclude air from the gel, but if not, then isobutanol is also used. The template may be readily eased out of a gel of about 10% T or less when it has fully polymerised. For gels of greater acrylamide content this may not be the case, but formation of the wells in another, narrow layer of low % T gel polymerised on top can overcome this problem. The rate of polymerisation is variable, according to the particular gel contents, but the process is usually complete within thirty minutes to two hours at room temperature.

When the gel has polymerised, persulphate and other contaminating ions remain which may slow down electrophoresis and which may interact with proteins and spoil their electrophoresis. These ions can be removed by electrophoresis (and replacement of reservoir buffers) *before* the sample is loaded ('pre-electrophoresis'). This process may be monitored by a fall in current (if constant voltage is used, or vice versa).

Although the system described above is probably the most common, alternatives exist to the radical-producing and crosslinking agents.[8,9] For instance, β-dimethylaminopropionitrile may be used instead of TEMED. In place of both persulphate *and* TEMED, riboflavin can be used. Its exposure to light in the presence of a *small* amount of oxygen (too much oxygen inhibits polymerisation) leads to the production of free radicals which initiate polymerisation. Riboflavin is used in small amounts (about 0.001% w/v) and does not interfere with electrophoresis. It is particularly useful in preparing gels used for isoelectric focusing where the presence of persulphate interferes with electrophoresis. 'Bis' is the most commonly-used crosslinking agent but the gel so formed is difficult to solubilise except by hydrogen peroxide treatments[10] or by commercial solubilisers. Solubilisation is desirable for scintillation counting of the weak β-emitters ^{14}C and ^{3}H, because any solid gel present can act as a quencher. Use of alternative, reversible, crosslinking agents can ease this problem. Several have been mentioned in the literature.[11,12]

4. Running the Gel

Various types of electrophoresis apparatus are available commercially, but others have been described in the literature and may be 'home-made' relatively cheaply. The basic design has two reservoirs of buffer (each containing an electrode) separated by the gel. The electrodes are connected to a direct current supply (commercially available) to provide the electric field. The sample is introduced to the system at the gel, usually at one end so that electrophoresis occurs through it. During prolonged electrophoresis the pH of the buffers in the two reservoirs may change as a result of electrolysis. To keep the reservoir buffers equal in pH it may be necessary to continually exchange them, and this may be done using an electric pump. Also during electrophoresis, heat is generated in the gel and this lowers the viscosity of the fluid medium and also the electrical resistance of the gel. Although this may accelerate the migration of proteins through the gel it can also increase the size of protein zones in the gel by diffusion, so causing loss of resolution. Cooling the gel can reduce this problem and is easily achieved by conducting the electrophoresis in a cold atmosphere, or by surrounding the gel (when in the apparatus) with one or both reservoir buffers, which are cooled. The buffer may be cooled by circulation through an ice bath, although excess cooling may precipitate some buffer components, such as sodium dodecylsulphate (SDS).

A typical slab gel apparatus, based on a design such as that described by Studier[13] is shown in Figure 2.2. In this design the gel is clamped against one side of the box which constitutes the bottom reservoir. One of the glass plates which supports the gel has a notch in it at the top (that is, a lower portion, in the middle) so that when the gel is in position buffer from the top reservoir can flow over at that point to make contact with the gel while *not* flowing right over the top of *both* plates into the bottom reservoir. In this design most of the gel is in contact with cooled buffer, although only on one side. An electrode of platinum wire runs along each of the reservoirs.

For horizontal slabs one side of the gel is exposed and electrodes are connected to each end via wicks of absorbent material dampened with buffer. Cooling may be provided by an underplate cooled by circulating water.

Before a sample is applied to the gel, it is dissolved in buffer. To ensure complete dissolution, loading buffers often contain agents such as SDS or urea which dissociate aggregates, and also reducing agents (e.g. dithiothreitol or 2-mercaptoethanol) which reduce disulphide bonds.

There are two common means of applying the sample solution to the gel. First, dissolved in the appropriate buffer, it may be spotted onto a piece of filter paper, which is then placed on the gel into which the proteins pass. The top reservoir buffer is best applied (carefully) after the sample. Secondly, the sample solution may be applied with a microsyringe, injecting it into a well in the gel. In this case the reservoir buffer is best applied *before* the sample, otherwise the inrushing buffer may disturb the sample. To aid this underlayering of the sample, the loading buffer often contains a solute which increases the density of the solution, such as sucrose (15% w/v) or glycerol (10% w/v). The act of loading is also easier if the sample solution is distinct from the reservoir buffer. It may be made so by inclusion of a suitable dye such as bromophenol blue in the loading buffer. Such a dye runs at or near the front of electrophoresis and can be used to monitor the progression of the run and possibly show up any erratic migration. It should not bind to the proteins.

The amount of sample loaded onto the gel is very variable, depending on the size of gel, the method used to detect the protein bands after electrophoresis (see Section 5), and the degree of resolution required, and is best determined empirically. However, in a suitable system 1 μg or less per protein band may be adequate.

As soon as the sample has been loaded onto the gel, electrophoresis is started by applying the electric field, which may be kept at constant voltage throughout electrophoresis or at constant current. As stated in Section 2, electrophoretic mobility is directly related to electric field strength. Thus, a high voltage will cause faster electrophoresis than will a low voltage, and so may avoid a little loss of resolution which would occur by diffusion of bands over a longer running time. However, it will also heat the gel more, which increases the rate of diffusion. Cooling the gel to reduce this problem slows the electrophoresis. The optimal conditions for electrophoresis are best determined empirically. Electrophoresis may be continued for as long as is desired but if no proteins are to be lost from the gel, it must be halted before the front (marked by the dye included in the sample application buffer) reaches the end of the gel.

When electrophoresis is complete, the gel is removed from the apparatus and treated to locate the protein bands on it (see Section 5). It is normal to do this immediately after electrophoresis so that slow diffusion does not blur the bands, but in practice this is generally not a great problem. Alternatively, the gel may be stored for later use, say for electrophoresis in a second dimension (Section 9). This may be done by immersion of the gel in suitable buffer followed by freezing in a moisture of solid CO_2 and methanol. The gel can then be stored at $-70°C$, and later thawed without any band diffusion or disruption of the polyacrylamide gel.

5. Location of Proteins on Gels

At some point it becomes necessary to locate the proteins on the gel. This is most commonly done by use of a dye which colours the protein bands but not the polyacrylamide gel itself. The literature describes a wide range of such dyes which are used in various mixes of solvents and the choice depends upon various points, such as:

(i) The dye must stain the protein in the type of electrophoresis system used. For instance, the presence of SDS may adversely affect the staining. Again, the background may stain, as can be the case with isoelectric focusing (Section 8) due to the background of ampholytes.

(ii) Sensitivity. It is usually desirable to use a sensitive stain, for more economical use of protein samples. One such dye is Coomassie Brilliant Blue R 250 (BBR 250), which binds by electrostatic forces to proteins, and which may be used in various mixtures of solvents. It can readily detect 1 μg of protein or less in a 5 mm-wide band in a 1 mm-thick gel although its efficacy can be somewhat variable, batch to batch. More sensitive staining may be achieved by following methods which employ the binding of Ag^+ ions to protein (e.g.[14]). While such methods are more sensitive than is Coomassie BBR 250 for many proteins we have found that the degree of difference varies from one protein to another and that some proteins do not stain at all by Ag^+ methods.

(iii) Purpose of staining. The amount of protein in a gel can be quantified by measuring the amount of dye bound to it. For this purpose it is necessary to use a dye which binds quantitatively to the protein, that is one for which the amount bound per μg of protein is constant over a useful range of protein weights. Not all dyes satisfy this condition. Since the amount of dye is usually estimated by the degree of its absorption at a particular wavelength of light, the uniformity of stain colour may also be important. For instance, Coomassie BBR 250 stains some proteins blue, others red (the so-called metachromatic effect). However, we have found that Procion Navy MXRB, which binds covalently to proteins, does not suffer in these respects, although it is somewhat less sensitive than Coomassie BBR 250.[15]

The staining procedure generally involves soaking the gel in the appropriate dye solution to allow it to diffuse into the gel and bind to the protein. Excess dye is then removed from those areas of the gel where it is not bound to protein by soaking in dye-free solvent mixtures (several changes). This works by diffusion, but destaining may be speeded up by placing the stained gel in an electric field to remove excess dye by its electrophoresis (sometimes called electrolytic destaining). Prolonged destaining may decolourise a protein stained with electrostatically-bound dye but if it is kept in solvents which fix or precipitate the protein the gel may be restained.

Various methods have been described which specifically locate certain types of proteins. For instance, glycoproteins can be detected by staining their carbohydrate component by the periodic acid – Schiff stain.[16] For enzymes, their catalytic activity may be used to locate them.[17] Again, proteins may be detected by the binding of specific agents such as antibodies. This can be done on the gel itself or on a replica of the gel made by blotting it (see Chapter 3). Lectins may also be used to detect glycoproteins by virtue of binding to their carbohydrate prosthetic groups (for instance, as in [18]). Detection of very small amounts of protein may be achieved if the protein (or antibody or any other agent used to detect it) is radioactively labelled, by scintillation counting of solubilised gel sections or by virtue of this radioactivity exposing a photographic film which is held against the gel (the techniques of autoradiography[19] and fluorography[20]).

It may be desirable to detect bands of protein without treating them in a way which irreversibly alters them, e.g. for preparative purposes or after the first stage of two-dimensional electrophoresis. This may be achieved by staining part of the gel and extrapolating to the rest, or by very brief and light staining (which leaves much of the protein unaffected). Other methods have been described for detection of unstained bands, for example using the phosphorescence of proteins under ultraviolet irradiation at low temperatures.[21]

Stained gels may be recorded by photography with bright-field illumination, using appropriate filters (such as yellow for obtaining best contrast with blue stained bands). Gels can be stored in buffers which do not promote destaining, or may be dried under vacuum onto supporting paper to prevent distortion of the gel.[22]

The following sections describe the more commonly used polyacrylamide gel electrophoresis systems.

6. SDS-polyacrylamide Gel Electrophoresis

SDS-gel electrophoresis is probably the most commonly used electrophoretic technique in protein chemistry. It gives information about the number and type of proteins present in a mixture, their relative abundance and a measure of their molecular weights. SDS (sodium dodecyl sulphate, $NaSO_3$-0-$C_{12}H_{25}$) is an anionic detergent that has a polar sulphate group linked to a non-polar aliphatic chain. It is reacted with the proteins before electrophoresis. Most proteins will bind about 1.4 g of SDS per gram of protein and the resulting denatured protein complex has been thought to approximate to a rod-like structure of uniform negative charge, though there are now thought to be regions that bind more detergent than others.[23] With the exception of a few structural proteins, protein-SDS complexes are soluble, and under the influence of an electrical field will migrate through a polyacrylamide gel towards the anode, generally at a rate inversely proportional to the logarithm of their molecular weight. This relationship may be used to determine the molecular weight of an unknown protein when run together with proteins of known molecular weight.

There are a number of systems for SDS-gel electrophoresis but the one developed by Laemmli[24] is the most widely used today, especially for slab gels. The gel is made in two parts. When using the apparatus shown in Figure 2.2, a 'running gel' or 'separating gel' of 14 × 12 × 0.1 cm is polymerised first. The buffer is Tris-HC1 pH 8.8 and contains SDS 0.1%

w/v. The concentration of the acrylamide and 'bis' can be varied from about 5% T for electrophoresis of proteins of 200,000 daltons to 20% T for low molecular weight proteins, say of 5000 daltons. Once this gel has polymerised, a 'stacking gel' is set on top of it. This is usually 2 cm deep and wells are formed in it with a suitable template. The buffer is Tris-HC1 pH 6.8, SDS 0.1% w/v. The acrylamide concentration is usually between 3–5% T and should be as low as possible to eliminate any sieving effect. The reservoir buffer is Tris-glycine pH 8.3, 0.1% w/v SDS. This buffer should not contain any other ions.

Proteins are dissolved in loading buffer, the composition of which is Tris-HC1 pH 6.8, containing 10% w/v glycerol or 15% w/v sucrose to make it dense, between 2 and 3% SDS and 0.01% w/v bromophenol blue. If the disulphide bonds of the protein are to be reduced, 2-mercaptoethanol is added to 5% v/v. It is important to make sure that the SDS has fully reacted with the protein and this can be done by heating to 100°C for three minutes or 37°C for one hour. Proteins that contain disulphide bonds will not be fully denatured by the SDS unless they are fully reduced.

Once the samples are loaded, electrophoresis is carried out with the cathode as the top electrode since SDS-protein complexes are all negatively charged. The reservoir buffers are circulated and cooled to about 10°C. A run may be carried out in three hours at 30 mA, after which time the bromophenol blue should have reached the bottom of the gel. At the end of the run the gel is removed and the proteins are detected either by staining or by some more appropriate method (see Section 5).

The 'stacking' system[25] is very useful for it leads to sharp, straight bands. The principle is known as isotachophoresis. The band-sharpening effect relies on the fact that the negatively charged glycinate ions (in the reservoir buffer) have a lower electrophoretic mobility than the protein-SDS complexes, which in turn have lower mobility than the Cl^- ions of the loading buffers and the stacking gel. When the current is turned on all the ionic species have to migrate at the same speed otherwise there would be a break in the electrical circuit. The glycinate ions can only move at the same speed as the Cl^- ions if they are in a region of higher field strength. Field strength is inversely proportional to conductivity which is proportional to concentration. The result is that the three species of interest adjust their concentrations so that $[Cl^-] > [protein\text{-}SDS] > [glycinate]$. There is only a small quantity of protein-SDS complexes so they concentrate in a very tight band between glycinate and Cl^- ion boundaries. Once the glycinate reaches the running gel it becomes more fully ionised in the higher pH environment and its mobility increases. Thus the interface between the glycinate and the Cl^- ions leaves behind the protein-SDS complexes which are left to electrophorese at their own rates.

An example of the separation that can be obtained by this technique is shown in Figure 2.3. The concentration of running gel (A) is 15% T and 2.67% C, and (B) is 7.5% T and 2.67% C. The stacking gel in both cases is

Figure 2.3: SDS Electrophoresis Carried Out in (A) a 15% and (B) a 7.5% Polyacrylamide Gel

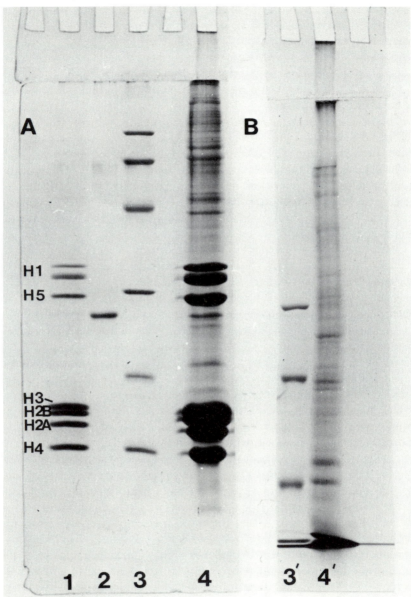

Note: Samples loaded. Lanes 1,4 and 4' an extract of chicken erythrocyte nuclei, 5,100 and 100 μg respectively. Lane 2 non-histone HMG 2, 1 μg. Lanes 3 and 3' molecular weight markers (Pharmacia), 5 μg. The molecular weights are, from top to bottom: Phosphorylase b 94,000; Albumin 67,000; Ovalbumin 43,000; Carbonic Anhydrase 30,000; Trypsin Inhibitor 20,000; ∝ Lactalbumin 14,400. All the samples were fully reduced and the gels stained in 0.1% Coomassie BBR 250.

4.5% T, 2.67% C. The figure shows a number of effects. Proteins of less than 30,000 daltons form SDS complexes which are barely retarded by the 7.5% gel and electrophorese at the ion front. Some proteins in lanes 4 and 4' are too large to enter the stacking gel. They would require a larger pore gel. Comparing lanes 1 and 4 one can see that overloading of samples makes the bands very broad but in this particular system they do not necessarily interfere with the lesser bands. The log of the molecular weights of the proteins in lanes 3 and 3' shows an approximately linear relationship when plotted against relative mobilities. However, it is not possible to correctly size all proteins with this method. For example, histones run anomalously – histone H4 (11,400 daltons) comigrates with $\propto$ lactalbumin (14,400 daltons). Thus, although the apparent molecular weight of a protein measured by SDS electrophoresis is a useful characteristic, when measured in this way it should not always be assumed to be an accurate measure of its true molecular weight. Indeed, there are examples showing that changing the amino acid sequence by just one residue can alter the apparent molecular weight by 2000 daltons.[26]

7. Other One-dimensional Electrophoresis Methods

There are several reasons why one may require alternatives to SDS electrophoresis. There are cases, for instance, when proteins will co-migrate on SDS gels, e.g. isoenzymes cannot be separated by this method unless they differ in molecular weight. In such cases alternative gel systems, which separate proteins according to differences in charge rather than in size, must be used. To ensure that the proteins have a high electrophoretic mobility in these systems, electrophoresis should be carried out at a pH several units different from their isoelectric points. The majority of proteins are anionic above pH 7. Thus they may be separated by electrophoresis on polyacrylamide gels in buffers above this pH. One of the first methods was described by Ornstein[27] and Davis.[28] They developed the discontinuous buffer system described in Section 6 (except for the inclusion of SDS) but many workers have found that it is not always necessary to use a stacking gel to obtain sharp bands. Thus Tris-glycine buffers at pH 8.3 or 9.5 have been used with the same buffer in the gel and the reservoirs. The 'trick' is to load the sample in a buffer which is 1/5 the concentration of that in the gel. This gives a band-sharpening effect by virtue of the low conductivity in the sample loading buffer, without using a discontinuous buffer system. These systems have been used in the analysis of serum proteins and a large number of other proteins. Enzymes may often be detected in such gels by their activity since they have not been denatured.[29] Electrophoresis conditions will vary with the apparatus used, but serum proteins can be resolved at 100V in about 4 hrs using 7% polyacrylamide gels. Histones and other basic proteins which are cationic at pH 9.5 will however run the 'wrong way' on the above systems. If the polarity of the electrodes is reversed the proteins will migrate into the gel but only slowly. In 0.9 M acetic acid, however, basic proteins are highly charged and can be separated by polyacrylamide gel electrophoresis more easily.[30] Several modifications of this method have been described which either improve or modify the separation that can be obtained.[31] For example acid gels may contain detergents to enhance separation. Histone H2B variants (subforms of the same protein that differ by only one or two amino acid residues) may be resolved if Triton X-100 is included in the gel. Thus this type of analysis is complementary to SDS electrophoresis, which would not normally separate these histone variants.

Another alternative system involves the use of polyacrylamide gradients. In principle gradient gels can be used with any buffer system and they do not require a stacking gel.[32] A gradient (linear or otherwise) can be set up between, say, 5–25% T. A protein loaded on the gel will migrate rapidly through the dilute (large pore) gel until it reaches a smaller pore-containing gel through which it can only move extremely slowly. Another protein with a lower charge, but of the same size and shape, will migrate more slowly but it will continue to do so until it too is virtually stopped at the same place by the pore size of the acrylamide gel. Thus the method separates largely on the basis of molecular weight. It is very useful when analysing mixtures of proteins of very different sizes on the same gel. For this reason SDS is commonly used in combination with gradient gels. The major problem is a technical one of obtaining a reproducible gradient, so for comparative purposes it is best to make a number of gels at a time. Gradient gels may be used to obtain estimates of molecular weights of proteins either with[33] or without SDS.[34]

8. Isoelectric Focusing

The isoelectric point (pI) of a protein is the pH at which the net charge of the protein is equal to zero. Substances such as proteins whose charge can vary in magnitude and polarity with pH are known as ampholytes. Commercial preparations of ampholytes can be obtained whose pIs vary from about pH 3 to 10. They are small molecular weight polyions which contain different mixtures of –COOH, –NH$_2$ or –SO$_3$H groups. When an electric field is applied to a mixture of ampholytes the ones with a low pI move towards the anode, whereas the ones with a high pI will migrate towards the cathode. Their rate of migration will be essentially dependent on their charge. However, with time they will sort themselves out in order of pI, and as they do so, their charge will diminish. This is because the pH in their immediate vicinity will approach their pI. When an ampholyte reaches a position where the pH is equal to the pI, it will no longer have any charge and so will stop migrating. When this is true for all the ampholytes in the system a pH gradient will have been set up. It is quite easy to see that if a mixture of proteins is introduced to the system each one would migrate to a position in the gradient where the pH was equal to its individual pI. This is known as isoelectric focusing. This technique is unlike the previous ones described for the resolution does not depend on the sieving effect of the polyacrylamide. The band width is independent of the sample size and once the proteins have reached their pIs they should not diffuse until the current is turned off.

Although tubes can be used, a horizontal slab or flat bed apparatus is the most convenient to use. The ampholytes are mixed with the acrylamide before it is polymerised. The polyacrylamide concentration should be very low (3–5%) to ensure free migration of proteins through the gel to their isoelectric points. The cathodic buffer is at a high pH, usually 1M sodium hydroxide. Conversely the anodic buffer is of a low pH usually 1M phosphoric acid. The apparatus should be cooled since the power used is high and when possible, isoelectric focusing should be run at constant power so that the heating effect is constant. In a typical run, the voltage may increase from 300 to 1200 volts and the current drops correspondingly as the ampholytes reach their isoelectric points. In general, bands are sharper the higher the voltage and the thinner the gel. Stable pH gradients can be achieved within one hour and provided that highly purified reagents are used the gradients will be stable for a few hours, although there may be a tendency for the ampholytes to drift to the cathode. This is mainly caused by electro-endosmosis (the migration of liquid caused by ions fixed on the polyacrylamide gel or the surface of the glass). More stable gradients can also be obtained by flushing the apparatus with N_2 to exclude CO_2. Samples are usually loaded on glass fibre paper onto the surface of the gel. Since proteins are often insoluble close to their pIs, they may never reach their true pI positions. This problem may be detected if the sample is loaded at both the anodic and cathodic ends of the gel. If the samples migrate to the same position then one is sure they have reached their pI. If not then urea and non-ionic detergents such as Nonidet P-40 can be added to the gel to overcome this problem.

A number of companies supply ampholytes and even ready-made gels in different pH ranges, from narrow ranges, over two pH units, to large range from pH 3–pH 9.5. Thus in 10 cm long gels it is possible to separate proteins whose pIs vary by less than 0.1 units. It is for this reason that it has become the method of choice for the analysis of isoenzymes and post-translational modifications of proteins. If required the precise pH gradient can be determined by cutting out 0.5 cm^2 pieces of gel, eluting them with degassed water (0.5 cm^3), and measuring the pH with a pH meter. Alternatively micro pH electrodes are available to measure the pH directly on the gel. It should be noted that the pI of proteins and the ampholytes vary with temperature and that this can lead to errors if the pH gradient is not determined at the same temperature as the separation.[35] Suitable pH markers may be run on the gel to define the gradient. Problems such as sample binding to ampholytes or chemical modification of amino acid side groups, which can cause artificial focusing patterns, are not uncommon. These problems are outweighed by the fact that an analysis can be carried out in two hours and that it is an extremely high-resolution technique. It is also one in which proteins are not usually irreversibly denatured and a large number of enzymes and isoenzymes are stable enough to be detected by their activity.

Staining of proteins on isoelectric focusing gels presents a special problem, for most protein stains will also stain the ampholytes. Early methods used a procedure that would fix the protein whilst washing out the ampholytes. A number of methods have now been described that do not require this step. One uses dilute copper sulphate solution to stabilise the protein-Coomassie BBR 250 complex whilst destaining the ampholytes.[36] Another which shows up the protein bands almost immediately uses Coomassie BBG 250 dissolved in perchloric acid.[37] An example of an isoelectric focusing gel run is shown in Figure 2.4. There are comprehensive reviews on the subject of isoelectric focusing in Righetti and Drysdale[38] and Smith.[39]

9. Two-dimensional Gel Electrophoresis

Although capable of excellent resolution the techniques mentioned so far are sometimes inadequate to resolve all the proteins in a complex mixture. However, by combining two electrophoretic techniques that separate proteins by different criteria, O'Farrell[40] has been able to resolve over one thousand protein spots on a two-dimensional gel slab. He separated the proteins by isoelectric focusing (i.e. on the basis of charge) followed by SDS-gel electrophoresis (i.e. on the basis of size). His excellent paper goes into great detail describing the apparatus, preparation of the gels and samples, and the electrophoresis conditions. Later, Anderson and Anderson[41] developed an apparatus that would take a large number of second dimension slab gels simultaneously which is an important factor when doing comparative work.

Briefly, the protein extracts are made in urea and non-ionic detergents. They are loaded on to the first dimension which is a 2.5 mm diameter isoelectric focusing rod gel that also contains urea, Nonidet P-40 and ampholytes. At the end of the run the gel is equilibrated with SDS gel-type sample buffer. It is then set on top of an SDS gradient slab gel using 1% (w/v) agarose dissolved in the same buffer. Bromophenol blue (0.001% w/v) is added to the top reservoir buffer. The buffers are not circulated for the first 30 min of the run by which time a sharp blue band should have formed in the stacking gel. Electrophoresis in a 0.1 x 14 x 14 cm slab should take about 5 hrs at 20 mA. The most abundant proteins may be visualised by staining, but the minor ones have to be radioactively labelled and detected by fluorography. In general, resolution is better when low quantities of protein

Figure 2.4: An Isoelectric Focusing Gel of Various Proteins using a pH 3 to 10 Gradient and Stained with Coomassie

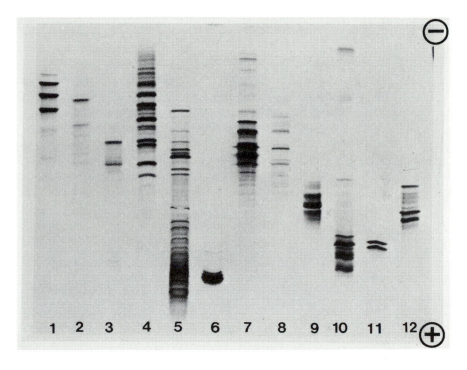

Note: Samples loaded: 1. Lens culinaris Lectin; 2. Myoglobin (sperm whale); 3. Myoglobin (horse heart); 4. Helix pomatia Lectin; 5. Helix pomatia blood; 6. Ovalbumin; 7. Plasminogen (porcine); 8. Plasminogen (human); 9. Soyabean Lectin; 10. Deoxyribonuclease I (bovine pancreas); 11. β-Lactoglobulin; 12. Conalbumin (chicken egg white).
Source: Pharmacia Fine Chemicals.

are loaded. This means that the proteins must be labelled to a high specific activity for optimal results.

The pH range that gives the greatest number of spots is pH 5–7. Of course not all proteins will focus in this range. A modified first dimension known as non-equilibrium isoelectric focusing[42] has been tried. It has the advantage of being able to map both acidic and basic proteins. If one is only interested in a particular group of proteins the pH range can be changed accordingly. The same is true for the second dimension – it often pays to use a non-gradient SDS gel which is much simpler to make reproducibly and often gives the necessary resolution.

The technique has been modified by a great number of workers. Basically any separation that can be carried out on a polyacrylamide gel may be used as the first dimension as long as it can be equilibrated with SDS sample buffer. Band diffusion during equilibration does not seem to be too much of

Table 2.1: Some Problems which can arise during Electrophoresis

Problem		Possible cause and solution
1. Failure/decreased rate of, gel polymerisation	a)	Oxygen present – degas.
	b)	Stock solutions aged – renew.
2. Poor gel top or well bottom	a)	Soft or sticky gel due to mixing of gel mixture with isobutanol or water during overlayering – overlayer *without* mixing.
	b)	Wells distorted or broken – remove well template carefully, or form wells in a lower % T gel layer.
	c)	Badly formed wells due to template fitting badly (loosely) – renew template.
3. Failure of proteins to enter gel	a)	Protein molecules too large – use gel of larger pore size (e.g. less acrylamide).
	b)	Proteins insoluble or aggregated – use reducing agents, urea, SDS or other detergents in buffers, where appropriate, or greater concentrations of these than previously.
4. Distortion of bands	a)	Proteins insoluble or aggregated – see 3 (b).
	b)	Insoluble matter or bubbles in gel – filter gel solutions and remove bubbles before polymerisation.
	c)	Inconsistency of pore size in gel – ensure gel solution is well mixed, and if polymerisation occurs too rapidly, slow it down (e.g. by cooling or reducing persulphate added). Remove bubbles – any rising through the gel mixture as it polymerises can introduce inconsistencies.
	d)	Insulation of part of gel – remove all bubbles adhering to the polymerised gel at the base before electrophoresis.
	e)	Electrical leakage along the sides of gel – insulate those parts properly.
	f)	Uneven cooling of gel allowing one part of the sample to run faster than others – improve cooling, reduce the thickness of the gel, or reducing heating by reducing voltage or ionic strength of buffers.
	g)	Uneven loading of sample – before loading check that the sample well bottom is flat and horizontal (see 2) and load evenly.
	h)	Sample is overloaded – repeat with smaller loadings.
	i)	Close proximity of other, heavy, bands of protein – repeat, using smaller loadings to reduce the

effect of the other bands, or change the system to alter their respective mobilities so that they no longer interfere with one another.

5.	Changes in electrophoretic mobility (generally relatively slight in the same system)	a)	Amounts loaded differ greatly – keep loadings approximately the same.
		b)	Constituents of gel variable in quality, batch to batch or with time – use one batch for as long as possible. Replace aged stock solutions.
6.	Contamination of samples on the gel	a)	Dirty apparatus (especially syringe used to load sample) – clean.
		b)	Handling of gel before final staining may deposit stainable material – wear gloves.
		c)	Sample solution may overflow into neighbouring wells – load as a more concentrated sample.
		d)	Seepage between layers of gel – when applying one gel to another wash isobutanol or water from first gel with appropriate buffer before applying the second gel, so as to get good adherence.
		e)	Stain may detect non-proteinaceous material in the sample (e.g. Procion Navy can stain DNA as well as protein) – try another stain.
7.	Unsatisfactory band detection (staining, autoradiography or fluorography)	a)	Weak staining – use a more sensitive stain, more concentrated dye in stain, or more prolonged staining (or exposure, in the cases of autoradiography and fluorography).
		b)	Uneven staining – increase staining time to allow dye to fully penetrate the gel.
		c)	Loss of stained bands – reduce destaining time. Ensure fixation of protein-dye complexes in destaining solvents.
		d)	Staining of background – use another stain (e.g. this problem may arise in isoelectric focusing – see Section 8), decrease background by prolonging the destaining process, and by frequently changing destaining buffer/solvents.
		e)	Very strong, perhaps indelible, non-specific marking of the gel by dye, due to solid particles of dye in stain-filter stain before use.
8.	Lack of resolution	a)	Volume of sample (zone size) is too great – load less sample, or more concentrated sample, or employ a suitable 'band-sharpening' system.
		b)	Bands have diffused – cool gel whilst running.
		c)	Bands are distorted – see 4.
		d)	Insufficient electrophoresis – prolong the run.

a problem. The fact that protein-SDS complexes are, in the main, larger than the native protein goes some way to slow down diffusion. One of the most important features of the second dimension is the stacking gel. Since the SDS complex takes time to electrophorese from the first dimension gel, this would lead to very diffuse spots on the second dimension gel were it not for the band-sharpening effect of the stacking gel.

Most gels can be equilibrated with SDS sample buffer within two hours, the exception being the acetic acid gels which require eight hours with four changes of buffer. Once equilibrated the first dimension gels may be stored at −70°C for months, without detriment. An autoradiograph of a typical two-dimensional gel is shown in Figure 2.5.

Figure 2.5: An Autoradiograph of a Two-dimensional Gel

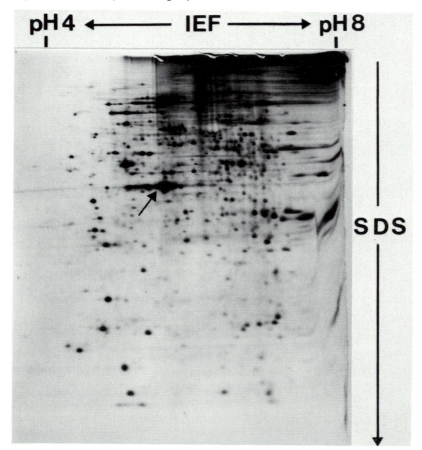

Note: An extract of rat tissue culture cells, labelled with ^{35}S methionine, was separated in the first dimension by isoelectric focusing (IEF) using pH 3.5 to 10 ampholytes. Electrophoresis in the second dimension was carried out using a 10% polyacrylamide SDS gel. The arrow indicates the position of actin.
Source: Dr S. Hobbs (with thanks).

10. Safety

Usual laboratory precautions should be employed during gel preparation and electrophoresis, but the neuro-toxicity of monomer acrylamide and 'bis' (by ingestion or skin contact) is worthy of note. *Poly*acrylamide is not toxic. The electrophoresis apparatus should be fully insulated and covered.

11. Problems

Good electrophoresis in polyacrylamide gels gives sharp, even, and well-resolved bands of protein. However, various problems can arise which prevent this. Some fairly general problems and their solutions are presented in Table 2.1. Problems associated with particular systems are not discussed, but as an example of one we may take the stain Coomassie BBR250, which can be used as a solution in methanol/glacial acetic acid/water (50:10:40 v/v) to stain proteins in SDS gels. This solid dye should first be dissolved in methanol *before* addition of the other components. We have found that otherwise its properties may be compromised – some proteins may not be stained at all if the solid is dissolved in the *full* mixture.

References

1. I. Smith, (ed.), Zone Electrophoresis, Chromatographic and Electrophoretic Techniques, Vol. 2 (W. Heinemann Medical Books Ltd, London, 1976).
2. A.H. Gordon, 'Electrophoresis of Proteins in Polyacrylamide and Starch Gels' in T.S. Work and E. Work (eds.), *Laboratory Techniques in Biochemistry and Molecular Biology* (North Holland Pub. Co., Amsterdam, London, 1969), vol. 1, 1–149.
3. C. Tanford, *Physical Chemistry of Macromolecules*, (J. Wiley and Sons, Inc., New York, London, 1961).
4. Gordon, *Electrophoresis of Proteins*.
5. Ibid.
6. D. Gross, 'Electrophoresis in Stabilizing Media. A Review', *Analyst, 90* (1956), 380–402.
7. F.W. Studier, 'Analysis of Bacteriophage T7 Early RNAs and Proteins on Slab Gels', *J. Mol. Biol.*, *79* (1973).
8. Smith, *Chromatographic and Electrophoretic Techniques*.
9. Gordon, *Electrophoresis of Proteins*.
10. E. Albanese and D. Goodman, 'A Simple Generally Applicable Method for Measuring Radioactive Substances Separated by Polyacrylamide and Agarose-Gel Electrophoresis', *Analyt. Biochem.*, *80* (1977), 60–69.
11. J. Bode, H. Schröter and K. Maas, 'Dissolvable Disulphide-Polyacrylamide Gels for the Electrophoretic Analysis of Chromosomal Proteins and for Affino-Electrophoresis of Thiol-Proteins', *J. Chromatog.*, *190* (1980), 437–444.
12. Smith, *Chromatographic and Electrophoretic Techniques*, 217.
13. Studier, 'Analysis of Bacteriophage T7', 237–248.
14. W. Wray, T. Boulikas, V.P. Wray and R. Hancock, 'Silver Staining of Proteins in Polyacrylamide Gels', *Anal. Biochem.*, *118* (1981), 197–203.
15. B.J. Smith, C.I.A. Toogood and E.W. Johns, 'Quantitative Staining of Submicrogram Amounts of Histone and High-Mobility Group Proteins on Sodium Dodecylsulphate – Polyacrylamide Gels', *J. Chromatog.*, *200* (1980), 200–205.
16. Smith, *Chromatographic and Electrophoretic Techniques*, 230.

17. M. Harris and D.A. Hopkinson, *Handbook of Enzyme Electrophoresis in Human Genetics* (North-Holland Pub. Co., Amsterdam, Oxford, 1976).
18. P.J. Robinson, F.G. Bull, B.H. Anderton and I.M. Roitt, 'Direct Autoradiographic Visualisation in SDS-Gels of Lectin-Binding Components of the Human Erythrocyte Membrane', *FEBS Lett*, *58* (1975) 330–333.
19. Gordon, 'Electrophoresis of Proteins'.
20. R.A. Laskey and A.D. Mills, 'Quantitative Film Detection of ^{3}H and ^{14}C in Polyacrylamide Gels by Fluorography' *Eur. J. Biochem.*, *56* (1975), 335–341.
21. J.K.W. Mardian and I. Isenberg, 'Preparative Gel Electrophoresis: Detection, Excision, and Elution of Protein Bands from Unstained Gels', *Anal., Biochem.*, *91* (1978), 1–12.
22. Smith, *Chromatographic and Electrophoretic Techniques*, 333–4.
23. C. Tanford, *The Hydrophobic Effect*, 2nd edn. (Wiley and Sons, Inc., New York, London, 1980).
24. U.K. Laemmli, 'Cleavage of Structural Proteins During the Assembly of the Head of Bacteriophage T4', *Nature 227* (1970), 680–685.
25. L. Ornstein, 'Disc electrophoresis-I. Background and theory', *Ann. NY Acad. Sci.*, *121* (1964), 321–349.
26. W.W. de Jong, A. Zweers and L.H. Cohen, 'Influence of Single Amino Acid Substitutions on Electrophoretic Mobility of SDS-Protein Complexes', *Biochem. Biophys. Res. Com.*, *82* (1978), 532–539.
27. Ornstein, 'Disc electrophoresis-I'.
28. B.J. Davis 'Disc Electrophoresis – II. Method and Application to Human Serum Proteins', *Ann. NY Acad. Sci.*, *121* (1964), 404–427.
29. Harris and Hopkinson, *Handbook of Enzyme Electrophoresis*.
30. E.W. Johns, 'The Electrophoresis of Histones in Polyacrylamide Gel and their Quantitative Determination', *Biochem. J.*, *140* (1967), 78–82.
31. R. Hardison and R. Chalkley, 'Polyacrylamide Gel Electrophoretic Fractionation of Histones', in G. Stein, J. Stein and L.J. Kleinsmith (eds.), *Methods in Cell Biology* (Academic Press, NY, San Francisco, London, 1978), chapter 17.
32. Smith, *Chromatographic and Electrophoretic Techniques*, chapter 10.
33. J.F. Poduslo and D. Rodbard, 'Molecular Weight Estimation Using SDS-Pore Gradient Electrophoresis', *Anal. Biochem.*, *101* (1980), 394–406.
34. P. Lambin and J.M. Fine, 'Molecular Weight Estimation by Electrophoresis in Linear Polyacrylamide Gradient Gels in the Absence of Denaturing Agents', *Anal. Biochem.*, *98* (1979), 160–168.
35. P.G. Righetti and T. Caravaggio, 'Isoelectric Points and Molecular Weights of Proteins, a Table', *J. Chromatog.*, *127* (1976), 1–28.
36. P.G. Righetti and J.W. Drysdale, 'Isoelectric Focusing in Gels', *J. Chromatog.*, *98* (1974), 271–321.
37. A.H. Reisner, P. Nemes and C. Bucholtz, 'The Use of Coomassie BBG 250 Perchloric Acid Solution for Staining in Electrophoresis and Isoelectric Focusing on Polyacrylamide Gels', *Anal. Biochem.*, *64* (1975), 509–516.
38. Righetti and Drysdale, 'Isoelectric Focusing'.
39. Smith, *Chromatographic and Electrophoretic Techniques*, chapter 11.
40. P.H. O'Farrell, 'High Resolution Two-Dimensional Electrophoresis of Proteins', *J. Biol. Chem.*, *250* (1975), 4007–4021.
41. N.G. Anderson and N.L. Anderson, 'Two-Dimensional Analysis of Serum and Tissue Proteins: Multiple Isoelectric Focusing', *Anal. Biochem.*, *85* (1978), 331–340.
42. P.Z. O'Farrell, H.M. Goodman and P.H. O'Farrell, 'High Resolution Two-Dimensional Electrophoresis of Basic as well as Acidic Proteins', *Cell*, *12* (1979), 1133–1142.

3 PROTEIN BLOTTING

Keith Gooderham

CONTENTS

1. Introduction

Polyacrylamide gel electrophoresis has proved to be an extremely powerful tool for the analysis of complex mixtures of proteins (see Chapter 2). However, while the resolving power of this method has progressively increased (the two-dimensional gel system of O'Farrell[1] for example being capable of resolving more than 1600 different proteins) any further analysis of the separated proteins has been limited. The majority of the protein molecules lie buried within the gel matrix and are therefore not readily accessible to further investigation. Though methods have been developed to overcome this problem (for example by elution of the proteins[2] or by direct *in-situ* analysis using antisera[3,4] or other protein probes[5]) these methods are in general time-consuming, insensitive and also frequently lead to a loss of resolution.

The realisation that methods essentially identical to the DNA blotting techniques pioneered by Southern[6] (see Chapter 15), could also be applied to proteins, has led to an important breakthrough in their analysis. Using the protein blotting method, the high resolution obtained by polyacrylamide gel electrophoresis is maintained but by transferring the proteins out of the gel matrix and onto the surface of a filter the fractionated proteins become accessible to analysis by a wide variety of probes. Over the last three years protein blotting has been applied to an increasingly diverse range of problems, including the identification of DNA and RNA binding proteins, glycoproteins and the screening of various antigens and antisera.

2. Methods

Transfer techniques

Three basic methods of protein blotting are currently used, capillary blotting, contact-diffusion blotting and electroblotting.

Capillary blotting of protein gels is performed essentially as described by Southern[7] for DNA gels (see Chapter 15). The advantages of this method are that the majority of the materials required will be readily available in most laboratories. Also several transfer experiments may easily be performed at any one time. The major disadvantage of the technique is that it is very slow, taking at least two days and even then only a partial transfer is generally achieved, with the high molecular weight proteins being particularly slow to leave the gel.

Contact-diffusion blotting[8] is a variation of the capillary blotting method and the transfer assembly is shown in Figure 3.1. Each layer of the sandwich is thoroughly pre-wetted with buffer and then carefully positioned on top of the previous layer, taking care to avoid trapping any air bubbles which would distort the resulting transfer. The assembled transfer-sandwich is totally immersed in buffer and the transfer then proceeds by diffusion. As with capillary blotting this method is very slow, requiring one and a half to two days and frequently longer in the author's experience. This method, like the capillary blotting procedure, has the advantage of not requiring the purchase or construction of any special apparatus. An additional advantage of this method is that two identical blots are produced, one of which can be used to provide a permanent record of the transfer while the other is challenged with the probe.

Figure 3.1: A Schematic Representation of the Protein Blotting Sandwich used in both Contact-diffusion and Electroblotting Experiments

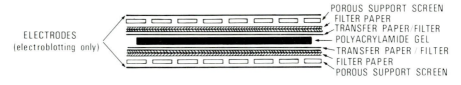

Note: In contact-diffusion blotting experiments protein migration is bidirectional and two transfer filters are required. Protein transfer is unidirectional in electroblotting experiments and a single transfer filter is used.

Electroblotting was first described by Towbin[9]. It utilises a similar sandwich arrangement to contact-diffusion blotting (Figure 3.1) but instead of being passively transferred the proteins are electrically driven out of the gel and onto the filter. This active transfer of protein molecules achieves much higher transfer efficiencies, together with shorter blotting times and the transfer conditions are precise and reproducible. The efficiency and length of transfer by this method are dependent upon the gel system used for the initial protein fractionation. Towbin[10] has shown that proteins from a polyacrylamide-urea gel can be completely transferred in as little as one hour. However, in order to obtain similar transfer efficiencies from polyacrylamide gels containing sodium dodecyl sulphate (SDS) Burnette[11] has reported that transfer times of twenty to twenty-four hours are required. In every case transfer is a function of molecular weight, with the largest proteins being transferred more slowly. Consequently the precise transfer conditions have to take into account the molecular weight range of the proteins under investigation and the composition of the gel. Where particularly large proteins (molecular weights in excess of 100,000 daltons) are to be transferred, blotting times can become prohibitively long. One way of overcoming this problem is first to break the proteins into a number of small peptides before transferring them from the gel. To this end Gibson[12] has developed a simple and ingenious *in-situ* digestion technique. The protein blotting sandwich is assembled as before, except that the filter papers on the cathode side of the gel are first soaked in a solution of Pronase (a cheap, stable, broad-spectrum protease). Once the electroblotting is started, the Pronase migrates out of the filter paper and into the gel where it is free to digest the proteins. The resulting peptides are then readily transferred out of the gel and onto the filter. An alternative to this method is to use the *in-situ* peptide mapping technique of Cleveland *et al.*[13] Though this method is comparatively more time consuming and restricted in the number of proteins which can be analysed at any one time, it does allow a detailed examination of the distribution of specific probe binding sites within the protein.[14] However, one problem which may arise with both of these methods is that while the protease digestion will considerably improve transfer efficiencies, it may at the same time destroy potential binding sites within the protein.

The major disadvantage of electroblotting is that, unlike the other two methods, it requires specific and potentially expensive apparatus. This method is therefore likely to be of greatest value where it is to be used regularly and where quantitative rather than qualitative transfers are required. Several companies, including Bio-Rad Laboratories, E-C Apparatus Corporation and Hoefer Scientific Instruments now manufacture electroblotting equipment. Alternatively a simple and very effective electroblotting apparatus can easily be made in the laboratory workshop.[15]

Transfer Media

Three principal types of immobilising matrices have been employed in protein blotting experiments: nitrocellulose filters, diazo papers and cyanogen bromide activated (CBA) papers.

Nitrocellulose filters are currently the most widely used transfer media and have the advantages of being relatively cheap, simple to use and they also have a long shelf life (at least twelve months at 4°C). These membranes are available from a number of manufacturers, including Bio-Rad Laboratories, Millipore (UK) Ltd and Schleicher and Schuell Inc. Parallel blotting experiments with DNA have indicated that pure nitrocellulose filters, rather than mixed ester filters, result in a more efficient binding of the DNA[16] and it seems likely that this is also the case for proteins. The precise nature of the protein-nitrocellulose binding is not as yet fully understood; however, it would seem to be the result of a number of different factors including hydrophobic and ionic interactions as well as hydrogen bonding. The binding capacity of the nitrocellulose filters is determined by the available surface area of the filter which is inversely proportional to pore size. Most workers have used 0.45 μm diameter pore filters; however,

Figure 3.2: The Major Steps in a typical Protein Blotting Experiment

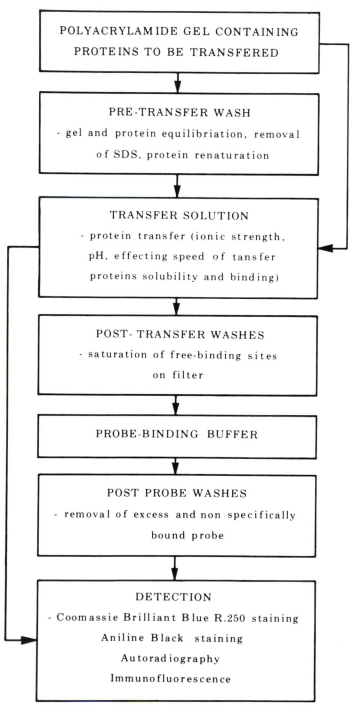

Burnette[17] has reported that smaller proteins are more efficiently bound when nitrocellulose with a 0.20 μm diameter pore size is used.

Two types of diazo papers are currently in use: diazobenzyloxymethyl (DBM) cellulose paper and diazophenylthioether (DPT) cellulose paper. As the reactive diazo groups are unstable these papers are initially prepared in a more stable intermediate form and then activated immediately prior to use. DBM papers can either be prepared in the laboratory by the method of Alwine[18] or more conveniently obtained from commercial sources in either the inactive nitrobenzyloxymethyl (NBM) cellulose form or the intermediate inactive aminobenzyloxymethyl (ABM) form. The NBM form has the advantage of having a longer shelf life (twelve months at 4°C) but must first be converted to the less stable ABM form (four months at 4°C under nitrogen, or longer at –20°C) prior to activation.

The DPT papers are similarly produced in an inactive form: aminophenylthioether (APT) cellulose paper. These papers are claimed by the manufacturer (Schleicher and Schuell Inc.) to be more stable than the ABM papers. The diazotisation step using acidic sodium nitrite is, however, identical for ABM and APT papers and in their activated forms they should possess similar binding affinities.

While the diazo papers require more preparation than nitrocellulose filters and must be used as soon as they are activated, they have the considerable advantage of covalently binding the transferred proteins to the paper. This permits the reuse of the protein transfer as many as four or five times after washing off the previous probe.[19,20] Owing to the short half-life of the diazonium groups on these papers, electroblotting, with its short transfer times, is the preferred method of blotting when using these papers.

CBA paper is not as yet commercially available but can be conveniently prepared according to the method of Clarke.[21] The transferred proteins are covalently bound to the CBA paper, though the precise mechanism of the reaction is as yet unclear. This paper therefore provides a cheap, though probably somewhat inferior, alternative to the diazo papers and has been successfully used by Bhullar and co-workers.[22]

A fourth class of transfer media, diethylaminoethyl (DEAE) anion-exchange membranes and filters (from Schleicher and Schuell Inc. and Bio-Rad Laboratories respectively) have also been used in a variety of DNA blotting experiments. However, there are as yet no published reports of their use in protein blotting experiments.

Transfer Buffers and Washes

The choice of transfer buffer and associated pre- and post-transfer washes is an important consideration in the design of any protein blotting experiment. The principal steps in a typical experiment are outlined in Figure 3.2. Though not all of these steps are necessary in every blotting experiment, the figure does draw attention to a number of important factors which must always be taken into account.

The inclusion of a pre-transfer step is frequently desirable in order to equilibrate the gel prior to transfer. The solution used is usually of a similar composition to that used for the transfer, in order to allow any changes in the size of the gel due to swelling or shrinking to occur at this stage rather than during the transfer. During this equilibration step 0.5% (v/v) Triton X-100 and 4 M urea have been used to promote the removal of SDS and to assist in the renaturation of the proteins.[23,24]

In order to obtain the most efficient transfer of proteins from the gel to the filter it is necessary not only that the blot is run for a sufficient length of time but also that the transfer buffer composition and pH will ensure the maximum solubility of all the proteins under investigation. In electroblotting experiments where Tris-glycine buffers are frequently used[25,26] an alternative buffer, for example the phosphate buffer used by Bittner,[27] should be substituted where the proteins are being transferred to either one of the diazo papers or CBA paper. This prevents the glycine from competing with the proteins for the available binding sites on the filter.

Upon completing the transfer, it is important to block any remaining free-binding sites on the filter in order to reduce non-specific binding of the probe, which would otherwise lead to a high background in subsequent steps. This neutralisation of free binding sites is generally achieved by incubating the transfer in either x1 Denhardt's reagent[28] or a 2–5% bovine serum albumin (fraction V) solution.[29,30,31]

The probe binding buffer should be designed to optimise the specific binding of the probe, while keeping non-specific binding to a minimum. For example, non-specific probe binding in eukaryotic protein-DNA binding experiments has been reduced by the addition of an excess of unlabelled prokaryote DNA.[32,33]

At the end of the probe incubation, excess unbound probe is removed by washing the filter. Non-specific probe binding has also been further reduced at this stage by sequential elution of the bound probe with increasing concentrations of salt.[34,35]

Protein and Probe Detection

An important step in any protein blotting experiment is to determine which proteins have been transferred to the filter. Although all of the proteins may have migrated out of the gel they need not necessarily have been retained by the filter. Also the proportions of the protein blot will often differ from those of a stained gel, the gel frequently being subject to changes in size during staining and destaining. Without prior visualisation of the transferred proteins these variations will therefore frequently make it difficult to relate the protein blot to either the standard gel picture or the probe binding pattern.

Detection of the transferred proteins may be achieved by staining with either Coomassie brilliant blue R 250[36] as in Figure 3.3 or with aniline blue black.[37] It is important to note that for the most efficient staining of the

Figure 3.3: Detection of DNA Binding Proteins

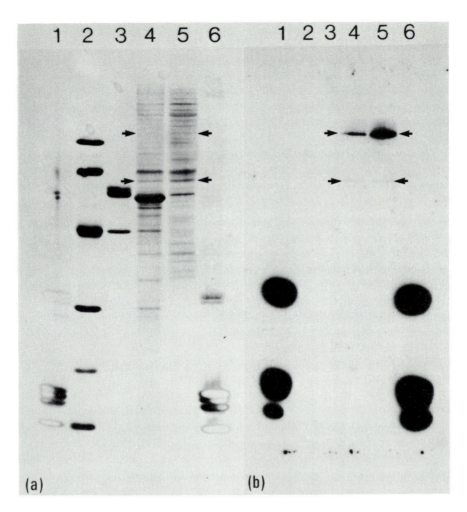

Note: Proteins were transferred to nitrocellulose by the electrophoretic method of Burnette[39] and detected either (a) directly by staining with Coomassie brilliant blue R 250, or (b) by autoradiography after incubation with a ^{32}P-labelled DNA probe and washing in 0.2 M NaCl as described by Bowen.[40] The protein samples were as follows: 1. total histones; 2. a molecular weight calibration mix (obtained from Pharmacia Fine Chemicals), composed of phosphorylase b, bovine serum albumin, ovalbumin, carbonic anhydrase, soybean trypsin inhibitor and α-lactalbumin; 3. tubulin (α and β subunits) and actin; 4. non-histone chromosomal protein fraction 1; 5. non-histone chromosomal protein fraction 2; 6. total histones. Note that only a partial transfer of the very basic histone proteins was achieved (tracks 1 and 6). Sufficient protein was, however, transferred to produce an extremely strong signal after incubation with the probe. Two non-histone chromosomal proteins (tracks 4 and 5, arrowed) also gave a strong signal when challenged with the probe. Both of these proteins are minor components in these samples but they clearly have a very high affinity for the DNA probe used in this experiment.

proteins the staining and destaining times given in these references should be followed precisely. Where a dry filter is to be stained it should first be soaked in water or it will severely distort, or even dissolve, in the high concentration methanol stain. Transfers which have been incubated in either Denhardt's reagent or bovine serum albumin will also generally give satisfactory results, though with a proportionately higher background.

Radioactively labelled proteins are usually avoided as they can interfere with the probe signal. They can, however, be used in some facilitated active transfer blotting experiments[38] (described in Section 3) when non-radioactively labelled probes are used (e.g. immunofluorescent probes).

Where radioactively labelled proteins are used, the filter can be dried and either auto-radiographed when using ^{125}I, ^{35}S and high activity ^{14}C as the label or fluorographed after treatment with En^{3}hance™ (New England Nuclear) for the detection of ^{14}C and ^{3}H.

3. Applications

Though the potential applications of protein blotting are extremely varied, they all share the same basic approach, i.e. a specific protein or group of proteins are recognised by their ability to bind a particular probe. The proteins may be challenged with the probe in either one of two ways. In the first and most frequently used method all of the proteins are transferred directly onto the filter and then incubated with the probe. In the second method the probe is initially bound to the filter and the proteins are then transferred and in this case only proteins recognised by the immobilised probe are retained by the filter. This latter method is known as facilitated active transfer. It was developed by Erlich[41] for use with DBM papers but should be applicable to other classes of transfer paper. The two techniques produce essentially identical results and the choice of which method to use will generally be a matter of experimental convenience. For example, experiments which involve the use of several different probes against a limited number of protein samples are more suited to the first method; whereas in experiments in which only one probe is being used to screen a large number of protein samples, the facilitated active transfer method will probably be favoured.

Protein blotting has found its widest application in the field of immunology as a rapid and sensitive screening technique for both antigens and antisera.[42,43,44,45,46] Unlike many previous methods, this technique does not require the initial purification or labelling of either the antigens or antibodies. Instead, the diluted antisera can be used directly and the resulting immune complexes then identified either with ^{125}I labelled *Staphylococcus aureus* protein A[47,48,49] or by using an appropriate anti-antibody labelled with ^{125}I[50,51] horse radish peroxidase[52] or fluorescein.[53] Using protein A as little as 0.1 ng of protein may be detected, though this will vary with the specific antibody titre.[54]

Another important application of the protein blotting methodology is in the detection of DNA and RNA binding proteins. Transferred proteins have been probed with radioactively labelled (^{125}I, ^{32}P or ^{3}H) DNA or RNA and the resulting nucleoprotein complexes detected by autoradiography or fluorography (see Figure 3.3 and examples given by Bowen *et al.*[55] and Gilmour *et al.*[56]). Using this method, specific DNA binding proteins and peptides, including both histone and non-histone proteins have been detected.[57]

Bowen[58] has also described the use of this method for the detection of specific protein:protein interactions, for example between the histones or various ribosomal proteins. Glycoproteins have also similarly been identified by Erlich[59] using lectin probes.

4. Conclusion

Though a number of practical problems still exist, for example in protein renaturation, these can largely be overcome by the use of either non-denaturing gel systems or by the insertion of pre- and post-transfer renaturation steps.

Protein blotting therefore offers an exciting new dimension to our work in protein analysis, considerably extending the information potential of polyacrylamide gel electrophoresis. The method is extremely versatile both with respect to the type of gel used for the initial protein separation and the variety of probes which can be employed. Also in addition to being a simple and highly sensitive analytical technique this method also clearly has preparative applications. Protein blotting has already become an important and routine tool in the field of immunology and will without doubt have a significant impact in many other areas of protein chemistry.

Further Reading

The development of protein blotting is still in its infancy and as yet, the most comprehensive and up-to-date source of further information concerning this technique comes from the various commerical companies in this field. For example:–

Schleicher and Schuell Inc., USA. Methods for the transfer of DNA, RNA and protein to nitrocellulose and diazotized paper solid supports. By: Barinaga, M. *et al.*, 1981.

and

Bio-Rad Laboratories. 'Southern, Western and Electroblotting', *Bulletin 1080 EG*, (June 1981).

References

1. P.H. O'Farrell and P.Z. O'Farrell, 'Two dimensional polyacrylamide gel electrophoretic fractionation', in G. Stein, J. Stein and L.J. Kleinsmith, *Chromatin and chromosomal protein research*, Methods in Cell Biology 16. (Academic Press, New York, London, Toronto, Sydney, San Francisco, 1977), Part 1, 407–420.
2. M. Bustin, R.B. Hopkins and I. Isenberg, 'Immunological relatedness of high mobility group chromosomal proteins from calf thymus', *J. Biol. Chem.*, *253* (1978), 1694–1699.
3. W.S. Adari, D. Jurivich and U.W. Goodenough, 'Localization of cellular antigens in sodium dodecyl sulphate-polyacrylamide gels', *J. Cell Biol.*, *79* (1978), 281–285.
4. K. Burridge, 'Changes in cellular glycoproteins after transformation: Identification of specific glycoproteins and antigens in sodium dodecyl sulphate gels', *Proc. Natl. Acad. Sci. USA*, *73* (1976), 4457–4461.
5. M.C. Snabes, A.E. Boyd III and J. Bryan, 'Detection of actin-binding proteins in human platelets by ^{125}I actin overlay of polyacrylamide gels', *J. Cell Biol.*, *90* (1981), 809–812.
6. E.M. Southern, 'Detection of specific sequences among DNA fragments separated by gel electrophoresis', *J. Mol. Biol.*, *98* (1975), 503–517.
7. Ibid.
8. B. Bowen, J. Steinberg, U.K. Laemmli and H. Weintraub, 'The detection of DNA-binding proteins by protein blotting', *Nucleic Acids Res.*, *8* (1980), 1–20.
9. H. Towbin, T. Stahelin and J. Gordon, 'Electrophoretic transfer of proteins from polyacrylamide gels to nitrocellulose sheets. Procedure and some applications', *Proc. Natl. Acad. Sci. USA*, *76* (1979), 4350–4354.
10. Ibid.
11. W.N. Burnett, '"Western blotting". Electrophoretic transfer of proteins from SDS-polyacrylamide gels to unmodified nitrocellulose and radiographic detection with antibody and radioiodinated protein A', *Anal. Biochem.*, *112* (1981), 195–203.
12. W. Gibson, 'Protease-facilitated transfer of high-molecular-weight proteins during electrotransfer to nitrocellulose', *Anal. Biochem.*, *118* (1981), 1–3.
13. D.W. Cleveland, S.G. Fischer, M.W. Kirschner and U.K. Laemmli, 'Peptide mapping by limited proteolysis in sodium dodecyl sulphate and analysis by gel electrophoresis', *J. Biol. Chem.*, *252* (1977), 1102–1106.
14. Bowen, Steinberg, Laemmli and Weintraub, 'The detection of DNA-binding proteins'.
15. M. Bittner, P. Kupferer and C.F. Morris, 'Electrophoretic transfer of proteins and nucleic acids from slab gels to diazobenzyloxymethyl cellulose or nitrocellulose sheets', *Anal. Biochem.*, *102* (1980), 459–471.
16. T.P. St John and R.W. Davis, 'Isolation of galactose-inducible DNA sequences from *Saccharomyces cerevisiae* by differential plaque filter hybridisation', *Cell*, *16* (1979), 443–452.
17. Burnette, "Western blotting".
18. J.C. Alwine, D.J. Kemp and G.R. Stark, 'Method for detection of specific RNAs in agarose gels by transfer to diazobenzyloxymethyl paper and hybridization with DNA probes', *Proc. Natl. Acad. Sci. USA*, *74* (1977), 5350–5354.
19. J. Renart, J. Reiser and G.R. Stark, 'Transfer of proteins from gels to diazobenzyloxymethyl paper and detection with antisera: A method for studying antibody specificity and antigen structure', *Proc. Natl. Acad. Sci. USA*, *76* (1979), 3116–3120.
20. J. Symington, M. Green and K. Brackman, 'Immunoautoradiographic detection of proteins after electrophoretic transfer from gels to diazo-paper: Analysis of adenovirus encoded proteins', *Proc. Natl. Acad. Sci. USA*, *78* (1981), 177–181.
21. L. Clarke, R. Hitzeman and J. Carbon, 'Selection of specific clones from colony banks by screening with radioactive antibody', *Methods in Enzymology*, (Academic Press, New York, London), *68* (1979), 436–442.
22. B.S. Bhullar, J. Hewitt and E.P.M. Candido, 'The large high mobility group proteins of rainbow trout are localized predominantly in the nucleus and nucleoli of a cultured trout cell line', *J. Biol. Chem.*, *256* (1981), 8801–8806.
23. Bowen, Steinberg, Laemmli and Weintraub, 'The detection of DNA-binding proteins'.
24. Towbin, Stahelin and Gordon, 'Electrophoretic transfer of proteins'.
25. Ibid.
26. Burnette, "Western blotting".

27. Bittner, Kupferer and Morris, 'Electrophoretic transfer of proteins'.
28. D. Denhardt, 'A membrane-filter technique for the detection of complementary DNA', *Biochem. Biophys. Res. Commun.*, *23* (1966), 641–646.
29. Towbin, Stahelin and Gordon, 'Electrophoretic transfer of proteins'.
30. Burnette, "Western blotting".
31. V. Shen, T.C. King, V. Kumar and B. Daugherty, 'Monoclonal antibodies to *Escherichia coli* 50S ribosomes', *Nucleic Acids Res.*, *8* (1980), 4639–4649.
32. Bowen, Steinberg, Laemmli and Weintraub, 'The detection of DNA-binding proteins'.
33. R.S. Gilmour, A. Lang, J.R. Gu, C. Johnston and J. Paul, 'DNA binding proteins in chromatin structure and function', *Cell Biol. Int. Repts.*, *5* (1981), 45–53.
34. Bowen, Steinberg, Laemmli and Weintraub, 'The detection of DNA-binding proteins'.
35. Gilmour, Lang, Gu, Johnston and Paul, 'DNA-binding proteins'.
36. Burnette, "Western blotting".
37. Bowen, Steinberg, Laemmli and Weintraub, 'The detection of DNA-binding proteins'.
38. H.A. Erlich, J.R. Levinson, S.N. Cohen and H.O. McDevitt, 'Filter affinity transfer. A new technique for the in-situ identification of proteins in gels', *J. Biol. Chem.*, *254* (1979), 12240–12247.
39. Burnette, "Western blotting".
40. Bowen, Steinberg, Laemmli and Weintraub, 'The detection of DNA-binding proteins'.
41. Erlich, Levinson, Cohen and McDevitt, 'Filter affinity transfer'.
42. Towbin, Stahelin and Gordon, 'Electrophoretic transfer of proteins'.
43. Burnette, '"Western blotting"'.
44. Symington, Green and Brackman, 'Immunoautoradiographic detection of proteins'.
45. Bhullar, Hewitt and Candido, 'The large high mobility group proteins'.
46. Erlich, Levinson, Cohen and McDevitt, 'Filter affinity transfer'.
47. Burnette, '"Western blotting"'.
48. Renart, Reiser and Stark, 'Transfer of proteins'.
49. Bhullar, Hewitt and Candido, 'The large high mobility group of proteins'.
50. Towbin, Stahelin and Gordon, 'Electrophoretic transfer of proteins'.
51. Erlich, Levinson, Cohen and McDevitt, 'Filter affinity transfer'.
52. Towbin, Stahelin and Gordon, 'Electrophoretic transfer of proteins'.
53. Ibid.
54. R.T.M.J. Vaessen, J. Kreike and G.S.P. Groot, 'Protein transfer to nitrocellulose filters', *FEBS Letts.*, *124* (1981), 193–196.
55. Bowen, Steinberg, Laemmli and Weintraub, 'The detection of DNA-binding proteins'.
56. Gilmour, Lang, Gu, Johnston and Paul, 'DNA-binding proteins'.
57. Bowen, Steinberg, Laemmli and Weintraub, 'The detection of DNA-binding proteins'.
58. Ibid.
59. Erlich, Levinson, Cohen and McDevitt, 'Filter affinity transfer'.

4 THE PRODUCTION AND ISOLATION OF PEPTIDES FOR SEQUENCE ANALYSIS

John M. Walker and Elaine L.V. Mayes

CONTENTS

1. Introduction and General Strategy

Although this chapter and Chapter 5 will consider the approaches necessary for determining the *total* amino acid sequence of a protein, it should be stressed that this is not always the ultimate aim of protein sequence studies. Determination of the *total* sequence of a protein is both a time-consuming (sometimes taking years) and expensive procedure, and such a project should not be taken upon lightly and without good reason. However, the trend in protein sequence analysis nowadays is not always towards the determination of *total* primary structures but more the rapid generation of *partial* sequence data sufficient to complement other observations. In recent years rapid developments in the fields of genetic engineering and DNA sequence analysis have made the direct sequencing of the gene for a given protein (and hence the deduction of its amino acid sequence) a viable alternative, in some cases, to the direct sequencing of a protein. It should be stressed, however, that some protein sequence data is still necessary for complete interpretation of the DNA sequence data. For example, comparison of peptide sequence data with DNA sequence data is necessary to identify initiation, termination and intron coding regions, or indeed the DNA coding region itself. Additionally, DNA sequence data provides no information whatsoever concerning the many protein modifications (e.g. acetylation, methylation, glycosylation, phosphorylation, cleavage of signal or pro-sequences etc.) that are known to occur as post-translational events, and these modifications can only be determined by peptide analysis. The interpretation of protein crystallographic data invariably requires some knowledge of peptide sequences, and partial sequences are sufficient to locate catalytic or binding sites and to indicate evolutionary divergence at these points. The requirement for peptide and protein sequence data is therefore as great as ever but the reasons for needing such data has changed considerably in recent years.

The basic strategy used in protein sequencing is easy to define. The native protein chain is broken down, either chemically or enzymatically, to give smaller polypeptides. As many as possible of these peptides are then isolated and their amino-acid sequences determined. Further native protein is then cleaved using a different cleavage method and the resultant peptides isolated and sequenced. In this way overlapping peptides are obtained which enable larger lengths of sequence to be deduced. In practice as many as half a dozen cleavage methods may be needed to produce all the necessary overlaps required to produce a completed structure. The choice of which type of cleavage method to use depends very much on the sequencing facilities available in any given laboratory. In general, enzymatic cleavages produce smaller peptides (five to 20 residues) suitable for manual sequencing, whereas the chemical methods tend to produce larger peptides (50–100 residues), which are more suited to sequenator analysis. (The distinction between manual and automated sequencing is described in Chapter 5). However prior to cleaving the protein a number of initial steps are necessary. These include:

(i) Assessment of the purity of the protein being studied. Ideally one should only commence protein sequencing when a protein is 100% pure. In practice, however, proteins containing as much as 10–15% impurity may be used, particularly if this 10–15% consists of a number of *different* proteins and not a single protein contaminant. When isolating protein cleavage products one generally only isolates the peptides which are produced in highest yield since peptide losses during peptide isolation are considerable. Low yield peptides from protein contaminants are therefore rarely isolated. Although there are obviously dangers with adopting this approach there are certainly occasions when this may be necessary. For example, one of the more practical problems with protein sequencing is producing sufficient protein (10–100 mg) to commence detailed sequence analysis. If protein preparative yields are low then a final purification step from 90% to 100% purity with associated loss in yield may not be acceptable or indeed advantageous. In this case it may be preferable to have sufficient protein for sequence analysis which is 90% pure, rather than insufficient protein that is 100% pure.

(ii) The molecular weight of the protein being studied should be determined. The molecular weight of the protein is an important factor since it allows the determination of the approximate number of amino acid residues in the protein, which in turn indicates the amount of effort that will be necessary to sequence the protein. In general, information on the protein molecular weight is obtained during its purification or the assessment of its purity.

(iii) If the protein consists of two or more polypeptide chains (sub-units) these chains should be separated prior to sequence analysis. The separated chains are then considered as individual sequencing projects.

(iv) The amino acid composition of the protein should be determined. This information, together with a knowledge of the molecular weight of the protein allows the determination of the approximate number of each amino acid in the protein. These figures are important in deciding which cleavage method to apply to the protein (see below).

(v) Any protein thiol groups should be blocked. The reduction of disulphide bridges (formed between cysteine residues, either as inter- or intra- chain linkages) and the blocking of the free thiol groups thus produced (or indeed any free thiol groups already existing in the protein) is a necessary step for any protein containing cysteine. If such groups are not blocked they are quite readily oxidised to disulphide bridges above neutral pH. Such oxidation can occur during peptide isolation steps and peptides thus linked are obviously of no use for sequence analysis.

Since all the above procedures involve standard methods in protein chemistry they will not be described in any detail here.

2. The Specific Cleavage of Proteins

Introduction

The specific cleavage of proteins may be achieved either chemically or enzymatically. Chemical cleavage methods aim to produce 100% cleavage at a small number of specific amino acids (usually methionine or tryptophan) and the most suitable reaction conditions are usually well documented. Proteolytic enzymes are also used to produce specific cleavages at given amino acids although the number of possible sites of cleavage are usually greater. However, in certain cases, proteolytic enzymes can be manipulated to produce more specific cleavages depending upon the conditions under which the reaction is carried out. In any enzymatic cleavage method, five main factors need to be taken into consideration. These are (a) pH; (b) temperature; (c) the protein to enzyme ratio; (d) the time of reaction and (e) the physical form of the protein being cleaved. The effect of each of these factors will now be considered.

The pH at which the reaction is carried out is a particularly important factor to take into account when using proteolytic enzymes to produce peptides. Generally the enzymes are used at their pH optima in order to produce maximum cleavage at each susceptible bond. However, the use of enzymes outside their pH optima can often result in a considerably reduced number of cleavage sites which gives rise to large peptides suitable for sequenator analysis.

The temperature, protein to enzyme ratio and the time of reaction all control the rate at which peptides are produced. A 'typical' cleavage reaction, for example, might be carried out at a protein to enzyme ratio of 200:1 at 37°C for four hours.

However, since all susceptible peptide bonds are not cleaved at the same rate by a given enzyme, greater enzymatic specificity can be introduced by reducing the rate at which the enzyme cleaves the protein and stopping the digestion at its early stages (digestion reactions may be stopped either by the addition of a specific enzyme inhibitor, adjustment of the pH to a value at which the protease is inactive, denaturation of the enzyme by pH changes or boiling, or freezing and lyophilisation of the sample). For example, the authors have used pepsin at a protein to enzyme ratio of 2,000:1 at 4°C for 20 minutes to produce cleavage at two specific points in a protein of 260 amino acid. Cleavage with pepsin on the same protein at a 200:1 ratio at 37°C for four hours produced over 50 small peptides.

Many proteins exist in highly folded (globular) structures. In order for enzymes and chemicals to gain access to amino acid residues 'buried' within these structures it is often necessary to denature the protein substrate, i.e. destroy its secondary and tertiary structures. Denaturation can be achieved by either boiling the protein solution, precipitation with strong acid (e.g. 10% trichloracetic acid), the use of detergents or the chemical disruption of disulphide bonds. Denatured proteins are often insoluble but when suspended in solution are suitable substrates for enzymatic cleavage. However it is not always necessary to denature globular proteins. Since such structures only allow proteolytic cleavage to occur (if at all) at the appropriate amino acid residues exposed at the surface of the structure, this can sometimes be used to our advantage since it limits the number of cleavage positions and may result in very few cleavages and thus production of specific large peptides.

For any cleavage method it is always necessary to carry out initial small-scale trial experiments to determine optimum cleavage conditions. This usually involves carrying out a series of time-course reactions at various pH values, enzyme to substrate ratios and temperatures. For the production of large peptides (> 50 residues) the course of any reaction can be followed by analysing samples by SDS gel electrophoresis. Final conditions should be chosen such that peptides are produced in high yield with little or no starting material remaining. The production of smaller peptides (five to 20 residues) is monitored by chromatography or electrophoresis (see below) and conditions usually chosen such that no further change occurs in the peptide pattern with time. This suggests that all susceptible bonds have been cleaved. The commonly used chemical and enzymatic methods for cleaving proteins are now described below.

Chemical Methods for Cleaving Proteins

In nearly all cases chemical cleavage methods do not produce 100% cleavage at the susceptible bonds. The pattern of peptides produced is therefore often somewhat more complicated than might be expected. For example, in a protein containing three susceptible bonds incomplete cleavage at each bond gives rise to nine peptides, whereas only four would be expected if cleavage were 100%. For proteins with a greater number of susceptible bonds the peptide pattern is correspondingly more complex. Despite this problem, if yields are sufficiently high then useful peptides can be recovered for sequence analysis. Indeed some peptides produced by incomplete cleavage may provide extremely useful information. The degree of cleavage at a particular bond varies considerably depending on the protein under investigation. For each method mentioned below there are individual cases where high cleavage yields have been achieved. Conversely there are proteins which are highly resistant to particular cleavage methods. It is therefore recommended that a variety of trial chemical cleavages are carried out on the protein being investigated to determine which methods and which cleavage conditions are best suited to that protein. The different chemical cleavage methods will now be discussed individually.

Cyanogen Bromide. Cleavage at methionine residues with cyanogen bromide (Figure 4.1) is probably the most successful of the chemical cleavage methods.[1] Cleavage occurs at the carboxyl side of methionine residues in high yields and since methionine residues are relatively uncommon in proteins, usually results in the production of large peptides suitable for sequenator analysis. The reaction is usually carried out in 70% formic acid at room temperature for 24–48 hours using a 100-fold excess of CNBr over methionine residues present. Prior to CNBr cleavage it is advisable to reduce the protein, since methionine residues can be oxidised to the sulphoxide or sulphone during protein isolation procedures. Such oxidised residues are not cleaved by CNBr.

Figure 4.1: Cyanogen Bromide Cleavage at Methionine Residues

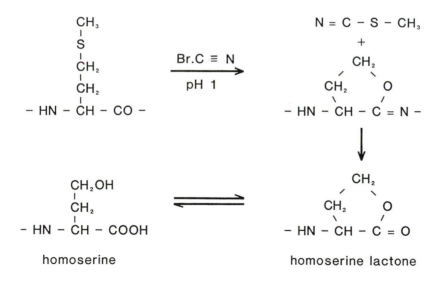

Following cleavage excess cyanogen bromide is removed (with care!) by lyophilisation, and the peptides fractionated. As a result of the cyanogen bromide cleavage reaction all methionine residues are converted to homoserine or homoserine lactone (which are in equilibrium with one another). Each peptide produced by cyanogen bromide cleavage therefore ends in a homoserine residue with the exception of the peptide derived from the C-terminus of the molecule.

Since the presence of homoserine in a peptide can be easily detected by amino acid analysis, any peptide that is isolated but does not contain homoserine must be the C-terminus peptide of the protein. It should be stressed that CNBr is a highly toxic and relatively volatile compound, and as such should be treated with respect. In particular great care should be taken to trap volatile cyanide products during the lyophilisation step.

BNPS-Skatole. BNPS Skatole (Figure 4.2) is a mild oxidising agent which causes cleavage at tryptophan residues.[2] The reaction is carried out in formic or acetic acid (50–70%) using a 10–100-fold excess of BNPS-Skatole over available tryptophan groups. Cleavage is carried out in the dark (BNPS-Skatole is light sensitive) for 12–48 hours at room temperature. Cleavage yields of 30–65% have been reported for this method. Since tryptophan residues usually only occur in very small numbers in proteins, often only once, this can be an extremely useful cleavage method. Side reactions that can occur include the oxidation of methionine residues and reaction at tyrosine residues. However, methionine can be regenerated from methionine sulphoxide at the end of the reaction by the addition of a reducing agent (e.g. mercaptoethanol) and reaction at tyrosine can be eliminated by carrying out the reaction in the presence of an excess of free tyrosine which acts as a scavenger. Specific modification of the tryptophan residue alone is therefore achieved.

Figure 4.2: BNPS-Skatole

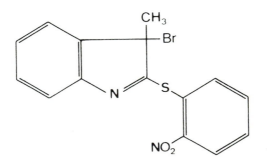

N-Bromosuccinimide. N-Bromosuccinimide (NBS) (Figure 4.3) is a much stronger oxidising agent than BNPS-Skatole and can produce cleavage at tryptophan, tyrosine and histidine residues.[3] However the rate of cleavage differs considerably at each residue the order being tryptophan > tyrosine > histidine. Therefore by using appropriate concentrations cleavage can be restricted to tryptophan residues. However since cleavage at tryptophan residues can be selectively achieved with BNPS-Skatole this method is probably most useful for cleaving at tyrosine residues in the absence of tryptophan or cleavage at histidine in the absence of both tryptophan and tyrosine. Reaction is carried out in acid media, usually in acetate or formate buffers below pH4, often in the presence of denaturing components (e.g. urea). A 2–10 molar excess of NBS is used and cleavage is carried out in the dark at room temperature for one to four hours. Cleavage yields are generally in the region 30–65% at each amino acid, although almost complete cleavage at tyrosine in histone Hl has been reported.[4] If N-chlorosuccinimide (NCS) is used cleavage is restricted to tryptophan alone[5] in yields up to 50% and as such NCS can be considered as an alternative to BNPS-Skatole.

Figure 4.3: N-Bromosuccinimide

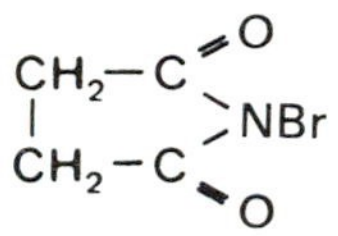

$$CH_2-C\overset{\displaystyle\nearrow O}{\underset{\displaystyle\searrow O}{>}}NBr$$

Other Chemical Methods. A number of other chemical cleavage methods have been reported in the literature in recent years and although not achieving the popularity of the above-mentioned reagents they have all proved useful in certain circumstances and as such should be considered when planning trial cleavage experiments:

(i) Two methods for specific cleavage at cysteine residues have been reported. Degani and Patchornik have shown that 2-nitro-5-thiocyanobenzoic acid converts cysteine residues to β-thiocyanoalanine.[6] Cleavage at residues modified in this way is then achieved by incubation at 37–50°C at high pH. This method has the disadvantage that peptides thus liberated have blocked N-terminal amino groups and therefore cannot be used for immediate sequence analysis. However, this method is useful for preparing peptide fragments for further cleavage with proteolytic enzymes. Doonan and Fahmy have produced specific cleavage at cysteine residues using an enzyme (from *A.mellea*) which specifically cleaves at the N-terminal side of lysine residues.[7] Protein lysine side chains are initially blocked by trifluoroacetylation and then cysteine residues subsequently converted to the 2-aminoethyl derivatives which are suitable substrates for the enzyme. Treatment of the modified protein with this enzyme therefore results in cleavage at the modified cysteine residues.

(ii) Treatment of proteins with hydroxylamine has been shown to cleave asn-gly linkages at yields up to 50%.[8] Most proteins can be reasonably expected to contain at least one such linkage and therefore the method should be included in trial digests.

(iii) Piszkiewicz *et al.* have reported that the asp-pro bond in proteins is particularly acid labile.[9] Since this observation a number of workers have exploited the lability of this bond by carrying out controlled mild acid hydrolysis of proteins to produce peptides. Since such a bond should not be very abundant in proteins, this method can be used to produce large peptides for sequence analysis.

(iv) In anhydrous acid, peptide bonds of serine (and to a lesser extent threonine) undergo intramolecular rearrangements with the side chain hydroxyl groups to form an ester link (an $N \rightarrow O$ acyl rearrangement).[10] This ester bond may then be cleaved under mildly basic conditions giving rise to cleavage at serine, and to a lesser extent threonine, residues.

Table 4.1: Proteolytic Enzymes of use for producing Peptides for Sequence Analysis

Enzyme	Cleavage site(s)	Optimal pH
1. Trypsin	C-terminal to Arg and Lys residues	8–9
2. V8 Protease (*S.aureus*)	C-terminal to Glu residues only in ammonium buffers C-terminal to Glu and Asp residues in other buffers	8 8 and 4
3. Clostripain	C-terminal to Arg residues	8
4. 'Lys' enzyme (*A.mellea*)	N-terminal to Lys residues	8
5. Chymotrypsin	Essentially C-terminal to Tyr, Phe, Trp and Leu residues but also to a lesser extent at Met, Asn, Gln, His and Thr	8–9
6. Pepsin	Essentially C-terminal to Phe, Tyr, Leu and Glu residues with many other minor cleavage positions	2
7. Thermolysin	N-terminal to all hydrophobic residues (e.g. Val, Leu, Ala, Ileu, Met, Phe)	8

Enzymatic Methods for Cleaving Proteins

Details of proteolytic enzymes that have been found to be useful in protein sequence determination are shown in Table 4.1. Ideally, to produce peptides suitable for sequence analysis proteolytic enzymes should produce a number of highly specific cleavages within the protein chain. The most useful enzymes for cleaving proteins are those which *specifically* cleave at only one or two different amino acids. In this respect trypsin has long proved useful in producing peptides since it produces cleavage C-terminal to only lysine and arginine residues, and even this specificity can be improved upon.

Modification of protein lysine side chains (e.g. by succinylation, see Figure 4.4) renders them resistant to tryptic cleavage and therefore trypsin cleaves only at arginine residues.[11] Recently three further proteolytic enzymes with high specificity have been introduced. V8-Protease (from *S.aureus*) cleaves essentially at glutamic acid residues in bicarbonate buffer (pH 7.8) or acetate buffer (pH 4.0) and at both aspartic and glutamic residues in phosphate buffer (pH 7.8) although occasionally other cleavage positions, especially at serine, have been reported.[12] The regular use of this enzyme in recent years has confirmed its usefulness in sequence studies. Clostripain is reported to cleave C-terminal to arginine residues only[13] but since this enzyme has only recently been made commercially available, its specificity and therefore its usefulness has not yet been comprehensively evaluated. Similar comment applies to the enzyme from *A.mellea* which is reported to cleave at lysine residues only.[14]

Figure 4.4: Succinylation of Lysine Side Chains

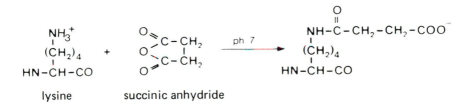

lysine   succinic anhydride

The other enzymes shown in Table 4.1 all have a reasonable degree of specificity, albeit not as great as those mentioned above, and have found considerable use in sequence analysis. Because of their reduced degree of specificity, peptide mixtures obtained from digests using these enzymes tend to be more complex. The usefulness of any one of these enzymes of course depends very much on the amino acid composition of the protein to be cleaved and this should be taken into consideration when choosing enzymes for trial digests.

Finally, certain other enzymes such as thrombin, plasmin, collagenase, etc. are often worth using in trial digests. Although these enzymes all have specific protein substrates *in vivo*, cleaving at highly specific peptide linkages within these proteins, they will sometimes also produce specific cleavages in proteins other than their natural substrates. Such cleavage positions are often quite novel and can provide most useful peptides.

3. The Assessment of Peptide Purity – the Dansyl Method for N-terminal Analysis

At many stages during peptide isolation it will be necessary to assess the purity of particular fractions, i.e. to determine whether further purification of the fraction is necessary. Although the presence of a single peptide band on chromatography or electrophoresis is indicative of a pure peptide, in general there is only one important criterion for defining a pure peptide – the presence of a single N-terminal amino acid. Fractions which contain more than one N-terminal amino acid are peptide mixtures and require further purification. A highly sensitive fluorescent method exists for the determination of N-terminal amino-acids. This is the dansyl method (Figure 4.5) first introduced by Gray.[15] An aliquot of the peptide solution (0.5n moles) is dried in a small tube (2″ × ¼″). To this dried sample 10 μl volumes of both sodium bicarbonate (0.1M) and dansyl chloride (2.5 mg/ml in acetone) are added and then the tube incubated at 37°C for one hour. During this incubation dansyl chloride reacts with the N-terminal amino group of the peptide (and any lysine side-chain groups present). Following incubation, the samples are dried under vacuum, 6N HCl (50 μl) added to the tube which is then sealed and heated overnight at 105°C. During this hydrolysis step the peptide is broken down into its component amino acids. However, the bond between the dansyl group and the N-terminal amino acid is stable under these conditions. This dansyl derivative of the N-terminal amino acid is highly fluorescent and can be identified by thin-layer chromatography. The hydrolysed sample is dried under vacuum, dissolved in 50% pyridine (10 μl) and 1 μl aliquots taken for chromatography. Chromatography is carried out on thin layer polyamide plates (5 x 5 cm). The sample alone is loaded on one side of the plate whereas on the reverse side the sample is loaded together with a mixture of a number of standard dansyl amino acis. The plates are then developed by chromatography in two dimensions. Dansyl amino acids are observed as highly fluorescent green spots when the plates are viewed under u.v. light. Two internal marker spots are also seen. These are dansyl amide (DNS-NH$_2$) formed by the reaction of dansyl chloride with traces of ammonia always present in solutions used, and dansyl hydroxide (DNS-OH, blue fluorescence) which is formed by hydrolysis of dansyl chloride. A reasonable idea of the identity of any unknown spot can be obtained by comparing its position (relative to Dansyl-hydroxide and amide) with a map of a chromatogram showing the relative positions of all the dansyl amino acids.[16] Confirmation of identity is obtained by observing the reverse side of the dansyl plate where the unknown is identified by its position relative to (or its superimposition upon) standard dansyl amino acids. The presence of a single fluorescent spot (other than DNS-NH$_2$ and DNS-OH) shows the peptide to be pure, and also allows the absolute identification of the N-terminal amino acid. This method is highly sensitive requiring less than 0.5n mole of a peptide and therefore makes no significant demand on one's peptide yield. One unavoidable disadvantage of this method is that dansyl derivatives of asparagine and glutamine are deaminated to the corresponding acids during the overnight acid hydrolysis step. It is therefore not possible to distinguish N-terminal aspartic acid from asparagine and N-terminal glutamic acid from glutamine. The use of the dansyl method in protein sequencing is described in Chapter 5.

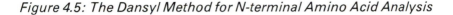

Figure 4.5: The Dansyl Method for N-terminal Amino Acid Analysis

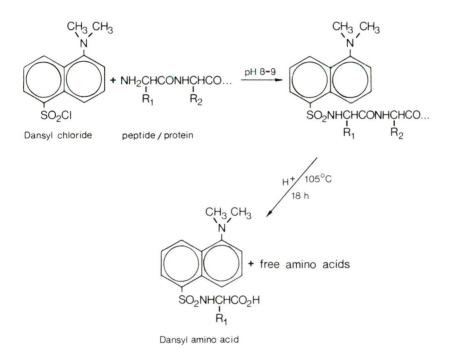

4. Peptide Purification Procedures

The purification of peptides is invariably the rate-limiting step in protein sequence determination. Most of the techniques that are available for peptide purification all involve relatively lengthy procedures and there has been little improvement in these techniques in the last few years. Fortunately, however, recent developments in the separation of peptides by HPLC (see below) are most encouraging and suggest that this may well become the method of choice for peptide separation in future years. Which approach to use when isolating peptides depends very much on the nature of the peptide mixture to be fractionated. Large peptides are generally produced in small numbers from any given digest and should normally pose no great purification problem. The methods used for isolating large peptides are essentially the same as those used in protein purification (mainly gel filtration and ion-exchange chromatography). Since a knowledge of protein purification techniques is assumed, the fractionation of large peptides will not be discussed further here. Conversely, protein digests that produce relatively small peptides (two to 20 residues) introduce the problem of separating a large number of peptides. The following discussion will therefore concentrate on techniques available for the separation of complex mixtures of small peptides. The main procedures in current use include gel filtration; ion-exchange chromatography; paper chromatography and electrophoresis, and high pressure liquid chromatography. Each method will now be discussed separately.

Gel Filtration

Gel filtration is generally used as a first step in peptide purification to produce an initial fractionation of the peptide mixture. Passage of a peptide mixture through a column of gel filtration media causes separation of the peptides according to their size, the peptides eluting from the bottom of the column in decreasing order of molecular weight. In practice a relatively crude fractionation into about half a dozen identifiable peaks is obtained. Generally, the larger the column used (1 m is a suitable length) the better the fractionation achieved. If only small columns are available a greater length may be obtained by connecting columns in series.

There is little limit to the quantity of sample that can be fractionated by gel filtration; the amount that can be fractionated depending mainly on the size (diameter) of the column available. Milligram quantities can be conveniently separated on columns of 1.0 cm diameter whereas gram quantities of material can be separated on columns in the 2.5–5.0 cm diameter range. The buffer used should be volatile, otherwise peptide fractions have to be desalted prior to further separation procedures. Acetic acid (5–50%), ammonium hydroxide (1M) or ammonium bicarbonate (1%) are therefore suitable solvents. The elution profile of peptide from the column may be determined by measuring the absorption of the eluant at 280 nm (measuring tyrosine and tryptophan) or 220–235 nm (measuring the peptide bond). Alternatively a small amount may be taken from each tube, subjected to alkaline hydrolysis and the amino acids produced detected by reaction with ninhydrin.[17] The intensity of the blue colour that develops is then recorded spectrophotometrically.

Having identified the elution profile the appropriate fractions are pooled and taken to dryness by rotary evaporation. These fractions can now be subjected to further purification procedures. There are two commercially available gel filtration media that are commonly used, namely Sephadex (a crosslinked dextran) and Bio-gel (a crosslinked acrylamide). Each is available in different grades according to the molecular weight range to be fractionated Sephadex G-25 (fractionation range 1,000–5,000) and Bio-gel

P6 (fractionation range 1,000–6,000) are generally the most useful for separating mixtures of small peptides. Peptide losses rarely amount to more than 10% following a gel filtration step.

Ion-exchange Chromatography

Like gel filtration, ion-exchange chromatography can also be used to produce an initial fractionation of a peptide mixture. However, since the resolving power is considerably superior to that of gel filtration some pure peptides may be obtained during such an initial fractionation. Chromatography of peptides can be carried out on either anion- or cation-exchange resins. Peptides are usually eluted by applying a linear pH and salt concentration gradient. To avoid the problem of having to desalt peptides, volatile components should again be used. In this respect pyridine-acetate or pyridine-formate mixtures are suitable since these are volatile and may be removed from peptide solutions by rotary evaporation or lyophilisation. A typical gradient might therefore be constructed from solutions of 0.1M pyridine-formate (pH 2.7) and 1.0M pyridine-formate (pH 6.5) for example.

Column elution profiles may be detected as described for gel filtration, although the use of these buffers excludes monitoring at wavelengths below 230 nm. Unfortunately, peptide losses are not insignificant with ion-exchange chromatography and may sometimes be as high as 30–40%. Following gradient elution it is always advisable to 'strip' the column resin by using either the appropriate extreme pH or a high salt concentration. In this way any peptides that are strongly bound to the resin and which would otherwise have been lost, are recovered. Frequently 30 or more peptide peaks can be produced from a protein digest fractionated by ion-exchange chromatography. Individual fractions are then dried by rotary evaporation or lyophilisation, redissolved in a small volume of water and a small aliquot taken for N-terminal analysis. Fractions showing a single N-terminal amino acid are retained for sequence analysis whereas mixtures are subjected to a further purification step.

Paper Chromatography and Electrophoresis

Since only a limited amount of peptide can be loaded onto any given chromatogram or electrophoretogram, the techniques of paper chromatography and paper electrophoresis are generally used following the initial fractionation of a peptide mixture. Chromatography and electrophoresis are carried out on sheets of suitably absorbant chromatography paper (e.g. Whatman No. 3MM). Samples to be separated should be applied at a concentration of no greater than 0.5 mg/cm since heavier loadings tend to cause distortion and streaking of peptide bands which obviously decrease the efficiency of the method. Chromatography and electrophoresis may be carried out using full size sheets (46 cm width) but sheets only a half or even a quarter this width may need to be used depending on the chromatographic or electrophoretic apparatus available. These techniques will now be discussed separately.

Paper Chromatography. For paper chromatography an aqueous solution of the sample to be fractionated is applied as a thin line along the width of a chromatography sheet approximately 10 cm from one end of the sheet and to within 2 cm of each edge of the sheet. The sample is applied with a microsyringe and dried in a stream of warm air. It is important that peptide samples are essentially salt free otherwise severe streaking of the chromatogram will occur. When all the sample has been dried onto the sheet the chromatogram is developed overnight by descending chromatography and then dried in a stream of warm air. Two different solvents that have found frequent use in peptide separation are shown in Table 4.2. To identify the position of the separated peptide bands 3 cm widths of chromatogram are cut from each side and stained for the presence of peptides (see below). These stained strips are then replaced adjacent to the main chromatogram and the position of the bands marked with a pencil. Peptides are recovered by cutting out the bands and eluting them into a suitable solvent (e.g. 5% acetic acid). Whether chromatography is an appropriate fractionation procedure to use for a given mixture, and if so which solvent to use, can be determined by initial trial chromatograms where only 1 cm widths of sample are run. In practice one would tend to analyse many fractions (each applied as 1 cm bands) in a single trial chromatogram.

Paper chromatography separates peptides up to approximately 20 residues in length, peptides much longer than this tending to remain at the

Table 4.2: Some Useful Buffers and Solvents for Paper Electrophoresis and Paper Chromatography

Method	Composition (v/v)	
Electrophoresis	pyridine : acetic acid : water (1:10:289)	pH 3.5
	pyridine : acetic acid : water (25:1:225)	pH 6.5
	pyridine : acetic acid : water (3:3:395)	pH 4.8
	formic acid : acetic acid : water (52:29:919)	pH 1.9
Chromatography	n-butanol : pyridine : acetic acid : water (15:10:3:12)	
	n-butanol : acetic acid : water (4:1:5) (upper phase only, lower in bottom of tank)	

origin. In general, the larger the peptide the smaller its mobility, and it may sometimes be necessary to run a chromatogram for a number of days in order to separate peptides with low mobilities. Chromatographic conditions should be chosen so that the majority of the length of the chromatography sheet is utilised to produce maximum peptide separation. Over 20 peptide bands can be identified in a single chromatographic run.

Paper Electrophoresis. Paper electrophoresis is somewhat more versatile than paper chromatography since one has a choice of pH at which to carry out the separation. Peptides which do not separate at one pH may be well separated at another and therefore trial separations should be carried out at a number of different pH values to determine the pH giving optimum separation. Some of the more routinely used buffers for peptide electrophoresis are shown in Table 4.2.

Samples to be separated are again loaded as a thin line across most of the width of the paper sheet. The sheet is then uniformly wetted with the appropriate buffer and electrophoresis carried out at 1000–2000 volts for one to two hours. Since at most pH values some peptides will migrate to the anode and some to the cathode, the position of application of the peptide sample, the voltage and the time of electrophoresis are all determined during trial runs and chosen to produce maximum separation within the length of electrophoretogram available. Electrophoresis is invariably carried out on a cooling plate since the considerable heat evolved during electrophoresis will distort the electrophoretogram unless dissipated. Peptide bands are identified and recovered as described for paper chromatography. Any peptides which do not separate at the chosen pH may be separated by carrying out a second electrophoresis at a different pH. Peptides up to approximately 20 residues in length can be separated by paper electrophoresis. Longer peptides tend to produce considerable streaking or if insoluble in the buffer used, remain at the origin. Like paper chromatography, 20 or more peptide bands can be separated in a single electrophoretic run. The combination of paper chromatography and paper electrophoresis therefore provides a powerful tool for fractionating peptide mixtures.

The Detection of Peptides on Chromatograms and Electrophoretograms

Side strips cut from chromatograms and electrophoretograms are generally stained by spraying with a ninhydrin solution (0.25% in acetone) and heating at 110°C for five minutes when peptides and amino acids appear as blue spots. Lines are then drawn across the chromatogram to connect the sides of these blue spots and the peptide bands cut out and eluted. The ninhydrin stain obviously identifies all possible peptides and amino acids in a given chromatogram. However, there also exists a number of stains which specifically identify peptides containing a particular amino acid. These include the Pauly stain for histidine and tyrosine,[18] the Sakaguchi[19] and phenanthrenequinone[20] stains for arginine and the Spies and Chambers stain for tryptotophan.[21] Peptides and free amino acids may be differentiated by the starch-iodine stain,[22] which stains peptide-bond containing material black, but produces no colour with free amino acids. These stains can be particularly useful when wishing to isolate certain specific peptides from a complex mixture. Such a situation can arise, for example, during the latter stages of protein sequence determination when only a small number of overlapping peptides (or even one) are required to complete the structure. Since the composition and possible structure of these peptides have invariably been deduced, if they contain an amino acid which can be specifically detected, such peptides can be relatively easily identified and isolated from a complex mixture.

Unfortunately, the use of stained side strips from chromatograms and electrophoretograms does not always result in the best possible recovering of peptides. This is particularly true for chromatograms since bands sometimes tend to run as 'wavy' lines and can 'bow' towards the middle of the chromatogram, particularly when using wide sheets. Cutting straight strips across the chromatogram can therefore miss a considerable amount of peptide. This problem may be overcome by lightly spraying the *whole* chromatogram with a solution of fluorescamine (0.05% in pyridine/acetone) when fluorescamine reacts with free amino groups to produce strong fluorescence under u.v. light (366 nm).[23] The whole peptide band can therefore be cut out and eluted accordingly. Since fluorescamine stained peptide is of no further use for sequence analysis, the strength of the fluorescamine solution is chosen such that only a small percentage of the peptide (sufficient for its identification under u.v.) is stained. Fluorescamine is considerably more sensitive than ninhydrin and may also of course be used to identify peptides on side strips. In this respect it is particularly useful for identifying weak peptide samples, being able to identify as little as 5 nmoles of peptide.

High Pressure Liquid Chromatography

With the recent developments in high performance liquid chromatography (HPLC) this technique is rapidly becoming one of the most powerful for purification of both proteins and peptides. HPLC offers considerable advantages, listed below, over the conventional techniques; its main disadvantage being cost.

(i) Higher yields. Recovery from HPLC is consistently high, usually greater than 90%, and in some cases as high as 100%. Although these yields are only a slight improvement on those of gel filtration columns, they are a considerable improvement on the yields obtained from both paper chromatography and electrophoresis, which are frequently as low as 20%.

(ii) Faster analysis time. A typical HPLC separation can be completed in one hour or less, whereas the conventional columns and paper chromatography and electrophoresis take at least one day, and frequently several days.

(iii) Higher resolution. The conditions for HPLC separation can be rapidly manipulated to give very well resolved peaks. An example of the exceptional resolving power of HPLC is shown by the reported separation of all the tryptic peptides obtained from lysozyme.[24]

(iv) Sensitivity. Very low amounts of peptides can be separated without large losses, as little as 10 ng on analytical columns. The sensitivity depends mainly on the sensitivity of the detection system. HPLC columns can be used over a large range of sample loadings – 10 ng–100 μg for analytical columns, and >10 μg for semi-preparative columns. Because of the short analysis times it poses no problem to the experimenter to run several HPLC analyses if larger amounts of peptides are to be separated.

For sequence analysis it is important to be able to remove the buffer by lyophilisation, in order to avoid further manipulation of the sample. Hence for peptide separation it is desirable to use volatile buffer systems such as ammonium acetate, ammonium bicarbonate, or trifluoroacetic acid. Unfortunately the best separation is not always achieved in these buffers and the experimenter may have to choose between better resolution with subsequent dialysis or re-chromatography of the sample in volatile buffer, and poorer resolution without subsequent manipulation of the sample.

The most sensitive means of monitoring the elution profile is by detecting the absorbance at 215 nm. If this is not possible then post-column derivatisation, such as with ninhydrin, or o-phthaldehyde, on aliquots of the fractions can be performed. Alternatively, the peptides can be succinylated or citraconylated before HPLC; this allows monitoring at 254 nm. As mentioned in Section 2 (Chemical Methods for Cleaving Proteins) these modifications can be used to limit tryptic digestion to arginine residues, or the modification can be performed subsequent to cleavage. This method has the added advantage that the peptides become more soluble in alkaline buffers, and the elution characteristics of reversed-phase HPLC may be improved.[25] If the citraconyl group is used then it can be removed at low pH after chromatography, but of course this will limit the HPLC elution buffers to neutral or alkaline pHs.

There are two types of HPLC normally used for peptide purification, gel permeation (or size-exclusion) and reversed-phase. The former separates peptides according to their molecular weight, the largest eluting first. However, under some circumstances charge and hydrophobicity can also affect elution time and this can sometimes be used to advantage. Reversed-phase HPLC separates peptides mainly on the basis of their hydrophobicity, the more hydrophobic peptides being retained longer on the column. Each method is discussed separately below.

Gel Permeation or Size-exclusion HPLC. Several of these types of columns are available commercially, those most frequently used for peptide separation are the I-125 columns from Waters Associates and the TSK-G 3000 SW type columns made by Toyo Soda and sold by several manufacturers. Both column types are suitable for separation of peptides greater than ten residues long. The resolution of these columns is improved by using denaturing buffers such as 6 M guanidine hydrochloride (pH 6.8), 6 M urea (in 0.2 M formic acid) or 0.1% SDS (pH 6.5).[26,27] However none of these buffers is volatile, hence the peptides obtained must be de-salted by either dialysis or reversed-phase HPLC. These gel permeation columns do not give good resolutions in the absence of salt. Separations in 1 M acetic acid, pH 3.0[28] and ammonium acetate, pH 7.0[29] do allow lyophilisation of the sample. Under these conditions peptides are not separated entirely on the basis of molecular weight, ionic and/or hydrophobic interactions also affecting the separation. Gel permeation HPLC is normally done at room temperature, with an optimum flow rate of 0.1 ml/min, although flow rates of up to 1.5 ml/min have been used.[30]

Reversed-phase HPLC. The most commonly used reversed-phase column supports used for peptide separations are octyl- (C_8-), octadecyl- (C_{18}-) and cyanopropyl- (CN-) silica. The C_8-columns are reported to give better resolution of larger peptides, and also have apparently longer column lives. Recently column packings with larger pore sizes have been available and these offer not only better separation of the smaller peptides but will also resolve larger peptides better than the smaller pore size columns.

When choosing the buffers there are several factors to be taken into consideration:

(i) As mentioned before the ideal buffers are volatile and permit monitoring at 215 nm. For these reasons ammonium acetate, ammonium bicarbonate and trifluoroacetic acid are amongst the most commonly used buffers.

(ii) The pH has a significant effect on resolution. Due to the lability of both the peptides and the silica particles only pHs between 2 and 8 may be used. Excellent resolution has been observed over the whole pH range.

(iii) The organic solvent most frequently used is acetonitrile, although methanol and n-propanol are also used (n.b. methanol must be driven off with a stream of nitrogen before lyophilisation). Peptides are usually eluted by linear gradients of the organic solvent, for example 5–30% acetonitrile[31] or 0–40% n-propanol.[32] Care must be taken that the higher concentrations of the organic solvents do not precipitate the peptides. For final purification the individual peptide peaks can be re-chromatographed under the appropriate isocratic conditions.

(iv) In the absence of inorganic modifiers peptides absorb to the stationary phase resulting in long retention times and broad peaks. The

Table 4.3: Conditions used for HPLC Separation of Peptides

Column	Buffer	Gradient	Flow rate	Sample	Reference
GEL PERMEATION					
LKB TSK-G 3000 SW	6 M urea/0.2 M formic acid or 0.25 M Na_2SO_4/0.25 M ammonium acetate pH 4.1 (adjusted with acetic acid)	—	1 ml/min	Various	33
Waters I-125	6 M urea/0.2 M formic acid	—	0.5 ml/min	CNBr[a] digests of haemagglutinin and tryptic digests of myoglobin	34
Waters I-125 and Beckman TSK-2000	1 M acetic acid pH 3.0 or 6 M urea/0.5 M acetic acid	—	0.5 ml/min	CNBr digests of actin and Spleen Ia antigen	35
REVERSED PHASE					
Various C_8 columns	0.1% TFA[b], 0.01 M HFBA[c] or 10 mM ammonium acetate, pH 6.5	5–30% acetonitrile	1–1.5 ml/min	Various	36, 37
Lichrosorb RP-18	pyridine (1 M)-acetic acid (0.5 M)	0–20% n-propanol	0.23 ml/min	Tryptic digests of ovalbumin & β-endorphin	38
Lichrosorb RP-8	0.013 M TFA	Various	0.7 ml/min	CNBr digest of haemoglobin	39
Zorbax CN	0.25 M phosphoric acid adjusted to pH 3 with triethylamine	0–40% acetonitrile	0.8 ml/min	Tryptic digests of neurophysins	40
Vydac TP 201 C-18	0.01 M HFBA	12.8–44.8% acetonitrile	1 ml/min	CNBr digests of collagen	41

Notes: a. CNBr = cyanogen bromide
b. TFA = trifluoroacetic acid
c. HFBA = heptafluorobutyric acid

higher the ionic strength (up to an optimum of 0.1 M) the narrower the peaks and the greater the column efficiency. 0.1% trifluoracetic acid is frequently used as the inorganic modifier since it is volatile, the pH is in a region where resolution is excellent (pH 2) and it also increases the solubility of the peptides in the organic solvent.

(v) Peptide separations by reversed-phase HPLC are normally done at room temperature with flow rates in the region of 1 ml/min.

In Table 4.3 some of the conditions used for separation of peptides have been summarised.

Further Reading

1.   S.B. Needleman (ed.), *Protein Sequence Determination* (Springer-Verlag, Berlin and New York, 1970).
2.   C.H.W. Hirs (ed.) *Methods in Enzymology* (Academic Press, New York) vols. 11 and 25B.

References

1.   E. Gross, 'The Cyanogen Bromide Reaction' in C.H.W. Hirs (ed.), *Methods in Enzymology*, (Academic Press, New York, 1967), 11, 238–255.
2.   A. Fontana, 'Modification of Tryptophan with BNPS-Skatole' in C.H.W. Hirs and S. Timasheff (eds.), *Methods in Enzymology* (Academic Press, New York, 1972), 25B, 419–423.
3.   L.K. Ramachandran and B. Witkop, 'N-Bromosuccinimide Cleavage of Peptides' in C.H.W. Hirs (ed.), *Methods in Enzymology* (Academic Press, New York, 1967), 11, 283–297.
4.   S.C. Roll and R.D. Cole, 'Quantitative Cleavage of Protein with N-Bromosuccinimide', *Journal of the American Chemical Society*, 92 (1970), 1800–01.
5.   M.A. Lischeve and M.T. Sung, 'Use of N-Chlorosuccinimide/Urea for the Selective Cleavage of Tryptophanyl Bonds in Proteins', *J. Biol. Chem.*, 254, (1977), 4976–80.
6.   Y. Degani and A. Patchornik, 'Cyanylation of Sulphydryl Groups by 2-Nitro-5-Thiocyanobenzoic acid. High Yield Modification and Cleavage of Peptides at Cysteine Residues', *Biochemistry*, 13, No. 1 (1974), 1–11.
7.   S. Doonan and H.M.A. Fahmy, 'Specific Enzymic Cleavage of Polypeptides at Cysteine Residues', *Eur. J. Biochem.*, 54 (1975), 421–226.
8.   M. Deselnicu, P.M. Lange and E. Heidemann, 'Studies on the Cleavage of the 2 Chains of Collagen with Hydroxylamine', *Hope-Zeyler's Z. Physiol. Chem.*, 354 (1973), 105–16.
9.   D. Piskiewicz, M. Landon and E.L. Smith, 'Anomalous Cleavage of Aspartyl-proline peptide bands during Amino Acid Sequence Determinations', *Biochem. Biophys. Res. Commun.*, 40 (1970), 1173–8.
10.   K. Iwai and A. Toshio, 'N→O Acyl arrangement', *Methods in Enzymology* (Academic Press, N.Y. 1967), 11, 263–82.
11.   I.M. Klotz, 'Succinylation', in C.H.W. Hirs (ed.), *Methods in Enzymology* (Academic Press, New York, 1967), 11, 576–80.
12.   G.R. Drapeau, Y. Boily and J. Houmard, 'Purification and Properties of an Extracellular Protease of Staphylococcus aureus', *J. Biol. Chem.*, 247, (1972), 6720–6.
13.   W.M. Mitchell and W.F. Harrington, 'Purification and Properties of Clostripeptidase B (Clostripain)', *J. Biol. Chem.*, 243, (1968), 4683–92.

14. R.A. Shipolini, G. L. Callewaert, R.C. Cottrell and C.A. Vernon, 'The Amino Acid Sequence of Phospholipase A_2 from Bee Venom', *Eur. J. Biochem.*, *48*, (1974), 465–76.
15. W.R. Gray, 'End Group Analysis Using Dansyl Chloride', *Methods in Enzymology*, (Academic Press, New York, 1972), 25B, 121–138.
16. Ibid.
17. C.H.W. Hirs, 'The Detection of Peptides by Chemical Methods', *Methods in Enzymology*, (1967), 11, 325–329.
18. S.B. Needleman, in *Protein Sequence Determination* (Springer-Verlag, Berlin and New York, 1970).
19. Ibid.
20. T. Yamoda, Y. Nakazawa and T. Ukita, *J. Biochem.* (Tokyo), 75 (1974), 153–164.
21. Needleman, *Protein Sequence Determination*.
22. Ibid.
23. E. Mendz and C.Y. Lai, 'Reaction of peptide with Fluorescamine on Paper After Chromatography or Electrophoresis', *Anal. Biochem.*, *65* (1975), 281–292.
24. M. Waterfield and G. Scrace, 'Peptide separation by high pressure liquid chromatography using size exclusion and reversed phase columns', in G.L. Hawk (ed.), *Chromatographic Science Series*, vol. 18 (Dekker, New York, 1982).
25. Ibid.
26. Ibid.
27. R.C. Montelaro, M. West and C.J. Issel, 'High performance gel permeation chromatography of proteins in denaturing solvents and its application to the analysis of enveloped virus polypeptides', *Anal. Biochem.*, *114* (1981), 398–406.
28. D.J. McKean, M. Bell, J. Moosic and B. Ballou, 'The separation and quantitative isolation of proteins and peptides by molecular sieve high pressure liquid chromatography' in G.L. Hawk (ed.), *Chromatographic Science Series*, vol. 18 (Dekker, New York, 1982).
29. R.A. Jenik and J.W. Porter, 'High-performance liquid chromatography of proteins by gel permeation chromatography', *Anal. Biochem.*, *111* (1981), 184–188.
30. M. Waterfield and G. Scrace 'Peptide separation by high pressure liquid chromatography using size exclusion and reversed phase columns'.
31. Ibid.
32. M. Rubinstein, 'Preparative high-performance liquid partition chromatography of proteins', *Anal. Biochem.*, *98* (1979), 1–7.
33. M. Waterfield, Personal communication (1982).
34. M. Waterfield and G. Scrace 'Peptide separation by high pressure liquid chromatography using size exclusion and reversed phase columns'.
35. McKean, Bell, Moosic and Ballou, 'The separation and quantitative isolation of proteins'.
36. M. Waterfield and G. Scrace 'Peptide separation by high pressure liquid chromatography using size exclusion and reversed phase columns'.
37. M. Waterfield, Personal communication.
38. M. Rubinstein, S. Chen-Kiang, S. Stein and S. Udenfriend, 'Characterisation of proteins and peptides by high-performance liquid chromatography and fluorescence monitoring of their tryptic digests', *Anal. Biochem.*, *95* (1979), 117–121.
39. W.C. Mahoney and M.A. Hermodson, 'Separation of large denatured peptides by reversed phase high performance liquid chromatography. Trifluoroacetic acid as a peptide solvent', *J. Biol. Chem.*, *255* (1980), 11199–11203.
40. I.M. Chaiken and C.J. Hough, 'Mapping and isolation of large peptide fragments from bovine neurophysins and biosynthetic neurophysin-containing species by high-performance liquid chromatography', *Anal. Biochem.*, *107* (1980), 11–16.
41. M. van der Rest, H.P.J. Bennett, S. Solomon and F.H. Glorieux, 'Separation of collagen cyanogen bromide-derived peptides by reversed-phase high-performance liquid chromatography', *Biochem. J.*, *191* (1980), 253–256.

5 PROTEIN AND PEPTIDE SEQUENCE DETERMINATION

John M. Walker

CONTENTS

1. The Edman Degradation

In 1950 Edman published a chemical method for the stepwise removal of amino acids from the N-terminus of a peptide.[1] This series of reactions has come to be known as the Edman Degradation, and although modifications of this technique have been introduced from time to time, (see below) the Edman Degradation remains, thirty years after its introduction, the only effective chemical means of removing amino acids in a stepwise fashion from a polypeptide chain. The overall reaction sequence is shown in Figure 5.1. The reactions may conveniently be divided into three stages:

(i) The Coupling Reaction: In this step the Edman reagent (phenylisothiocyanate, PITC) reacts with the amino group of the N-terminal amino acid of the polypeptide chain to give the phenylthiocarbamyl (PTC) derivative of the peptide. The reaction is carried out in an inert atmosphere of nitrogen. (Amino side chain groups of lysine residues within the polypeptide chain also couple with PITC at this stage, but this is not relevant to the Edman degradation.) The reaction takes place at pH 9.0–9.5, (the α-amino group must be unprotonated for reaction with PITC) at 50°C (20–40 minutes), usually in a volatile buffer. At the end of the coupling reaction volatile buffers and excess PITC are removed under vacuum. The PTC derivative is then washed exhaustively with organic solvents, usually benzene and ethyl acetate, to extract non-volatile buffers (if used) and in particular non-volatile side-products produced during the coupling reaction. These side products are mainly diphenylthiourea (formed by the hydrolysis of PITC) and phenylthiourea (formed by the reaction of PITC with ammonia). The washed PTC derivative is finally dried under vacuum.

(ii) The Cleavage Reaction: In this step the dried PTC derivative is treated with anhydrous acid (e.g. trifluoroacetic acid or heptafluorobutyric acid) at 50°C for 10–15 minutes. This rapid, non-hydrolytic step results in the cleavage of the PTC-polypeptide at the peptide bond nearest to the PTC substituent, thus releasing the original N-terminal amino acid as the 2-anilino-5-thiazolinone derivative, leaving the original polypeptide chain less its N-terminal amino acid. It is particularly important at this step that no traces of water are present. If water is present then the polypeptide chain will be exposed to hydrolytic conditions which will cause a small degree of internal cleavages within the polypeptide chain. Such newly liberated N-terminal groups have the disadvantageous effect of increasing the general

Figure 5.1: The Edman Degradation.
(I) The Coupling Reaction; (II) The Cleavage Reaction; (III) The Conversion Reaction.

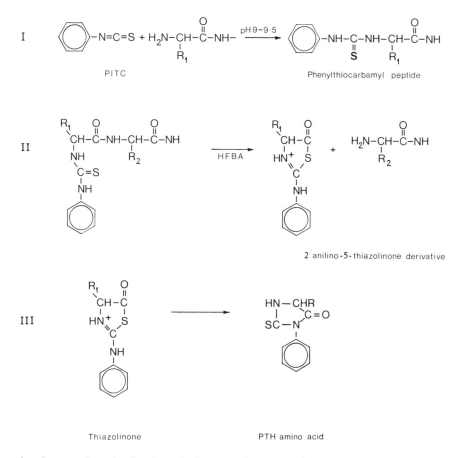

background at the final analysis stage (see below). Following the cleavage reaction the anhydrous acid is removed under vacuum and the thiazoline derivative extracted from the remaining peptide with an organic solvent (e.g. butyl acetate, butyl chloride), and recovered by removal of the organic solvent under a stream of nitrogen.

(iii) The Conversion Reaction: Since the thiazoline is a derivative of the N-terminal amino acid it may, in principle, be used for identification. However, in practice this is not done, since the thiazalinones are relatively unstable. A more stable derivative is obtained by converting the thiazolinone to the isomeric 3-phenyl-2-thiohydantoin (PTH) derivative. This is done by heating the thiozolinone in 1 N HCl at 80°C for ten minutes. The PTH amino acid is therefore the end-product of one cycle of the Edman degradation and requires identification.

The remaining polypeptide chain may now be subjected to further cycles of the Edman degradation. At the end of each cycle a PTH amino acid is recovered and identified. In this way the amino acid sequence of a polypeptide chain may be determined.

2. Identification of PTH Amino Acids

Three methods are routinely used to identify PTH amino acids. In practice one would use at least two methods to identify any given PTH sample since for each method there are always one or two PTH derivatives which are poorly resolved. Confirmation of one's results by a second method is therefore often required.

Identification by HPLC

With the increasing popularity of HPLC this method is rapidly becoming the first method of choice for identifying PTH amino acids. Not surprisingly, a number of different separation procedures have been published. One used routinely involves separation by reverse-phase chromatography on an ODS column using an acetonitrile gradient (25–40%) in a sodium acetate (0.01N, pH 4.5) buffer.[2] The PTH amino acids are detected by their u.v. absorbance at 269 nm and characterised by their retention times. Separation by HPLC has the advantage of being quick (10–15 minutes per sample); extremely sensitive (can identify as little as 50 p moles); quantitative, and in particular, has the potential for automation. This is particularly useful when a large number of derivatives are being produced each day (e.g. from an automated sequenator). Prior to the introduction of HPLC, PTH amino acids were often identified by gas chromatography. However, since gas chromatography required derivatisation of the PTH amino acid, the direct identification of PTH amino acids by HPLC is preferred nowadays.

Identification by Thin-layer Chromatography

PTH amino acids may be resolved by one-dimensional TLC. This is usually carried out on silica gel plates containing a fluorescent indicator. Following chromatography, PTH amino acids are identified by viewing under u.v. light when they appear as dark spots on a bright fluorescent background. A variety of different organic solvent systems are available for the separation of PTH amino acids[3] but all take approximately two to three hours for a plate to run. An advantage of the TLC method is that as many as ten samples may be run on a single plate (together with standard PTH amino acid samples). A sequence is immediately apparent by observing, on each track, the appearance of the new PTH amino acid (and its disappearance at the next step). A typical chromatogram is shown in Figure 5.2. A further advantage of the TLC method is that many volatile side-products may be removed by heating the developed chromatogram in an oven at 120°C for ten minutes. This gives a much 'cleaner' chromatogram and affords easier identification of certain PTH amino acids. Additional or confirmational information may be obtained from a TLC run by spraying the plate with a ninhydrin-collidine mixture and heating.[4] Many PTH amino acids produce characteristic colours with this stain and these colours aid in their identification. Recent developments in the chromatographic separation of PTH amino acids includes chromatography on micropolyamide plates (5 × 5 cm) in both one and two dimensions. Detection limits of down to 0.05 n moles are claimed for these methods with chromatography times of 30 minutes.[5] One disadvantage of all TLC methods is the fact that they are non-quantitative. They are, however, cheap and easy to set up, requiring no sophisticated machinery.

Figure 5.2: Thin-layer Chromatogram of the First Eight PTH Derivatives produced by Automated (spinning cup) Sequencer Analysis of a 90 Residue Peptide

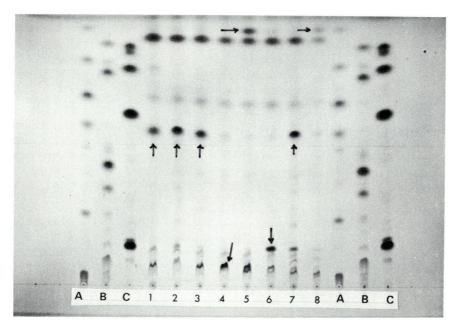

Note: Tracks marked A, B and C are different mixtures of standard PTH amino acids.
     The sequence identified in tracks 1 to 8 is: Gly-Lys-Gly-Asp-Pro-Asn-Lys-Pro. The strong spot appearing near the top of all the sequenator samples chromatographed is diphenylthiourea. PTH lysine and PTH glycine residues run in similar positions in the solvent system used but are easily differentiated using the ninhydrin-collidine stain where glycine gives a bright orange colour and lysine a dark blue colour. Similarly PTH aspartic acid tends to give a distorted spot in a region of the chromatogram that usually contains some background spots, but again this residue can be easily confirmed with the ninhydrin-collidine stain.

Source: Dr K. Gooderham (with grateful acknowledgement.)

Identification by Back-hydrolysis to the Free Amino Acid

PTH amino acids can be hydrolysed in strong acid (e.g. 57% hydriodic acid) to release the free amino acid. This method is referred to as 'back-hydrolysis', and the liberated amino acid can be identified by amino acid analysis.[6] The method is semi-quantitative, in that not all amino acids are recovered in 100% yield. However, knowledge of recovery rates from standard PTH amino acids allows hydrolytic losses to be accounted for. Also some amino acids produce readily identifiable degradation products (e.g. PTH threonine gives α-aminobutyric acid). This is a particularly useful technique in that any laboratory carrying out sequence analysis will amost certainly be equipped with an amino acid analyser. One major disadvantage of this method, however, is that derivatives of tryptophan, cysteine and serine are destroyed and asparagine and glutamine are converted to the corresponding acids. These observations emphasise the necessity of using at least two methods for PTH amino acid identification.

3. Repetitive Yield in the Edman Degradation

Theoretically, given a purified polypeptide, one should be able to apply a series of Edman degradation reactions and obtain the complete sequence of any polypeptide (even a long protein). In practice this is not possible since the Edman degradation does not give a 100% repetitive yield at each step. The repetitive yield at each step is an extremely important factor since it determines the amount of sequence data obtained in any sequencing procedure. As Table 5.1 shows, even a 1% difference in the repetitive yield has a great effect on the amount of data obtained (such figures are obtained by assuming that when the overall yield falls to 30% no more useful information is obtained since at this level the PTH amino acids cannot be identified against the background). The fact that the repetitive yield is always less than 100% is due to a number of factors; these include:

(i) Oxidative desulphurisation of the phenylthiocarbamyl group. The sulphur atom in the PTC derivative (Figure 5.1) is readily oxidised by traces of oxygen or peroxides dissolved in the buffer medium. The resulting phenylcarbamyl group acts as a blockage to the degradation since no thiazolinone can be formed. Repetitive yields are thus reduced.

(ii) The N-terminal amino group of the polypeptide chain can be blocked by traces of aldehydes (which are ubiquitous) which form Schiff bases with the N-terminal amino group. This reduces the repetitive yields.

(iii) The polypeptide is often mildly soluble in the organic solvents used for the washing and extraction steps. The amount of polypeptide therefore decreases at each cycle and consequently less PTH amino acid is available for identification at each step. This problem becomes more serious as the length of the polypeptide chain is reduced since the shorter the peptide the greater its solubility.

(iv) Many proteins become insoluble following one cycle of the Edman degradation. Coupling with PITC often causes insolubility as does the denaturing conditions of the cleavage step. 100% reaction at each step is highly unlikely with an insoluble protein, particularly if it is sitting as a 'lump' in the bottom of the tube! Similar insolubility problems do not usually exist for peptides. The introduction of automated sequence methods (see below) have largely managed to overcome these problems.

Table 5.1: The Relationship between Repetitive Yields in the Edman Degradation and the Number of Useful Cycles Achieved

Repetitive Yield	Number of Cycles
96%	25
97%	40
98%	60
99%	120

4. Manual Sequence Analysis

The Dansyl-Edman Technique

The carrying out of a series of Edman degradations on a sample in a test tube is referred to as 'manual sequencing'. Manual sequencing of total proteins (or large peptides) rarely gives information on more than the first two or three residues. This is generally due to the insolubility problem described in Section 3(iv). Extensive sequence data for native proteins and large peptides is best obtained by using an automated sequencer which is designed to overcome this problem (Section 5). Manual sequencing is generally carried out on peptides in the range five to 30 amino acids long, where the repetitive yield obtained by manual sequencing is typically 90–95%. The loss in yield is mostly due to oxidative desulphurisation caused by small amounts of oxygen which are difficult to exclude in the manual technique. Such repetitive yields generally allow the determination of sequences up to 15 residues in length, but in favourable circumstances somewhat longer sequences have been determined. Originally peptide sequences derived by manual sequencing were determined by identification of the released PTH amino acid. However in 1970 Hartley described the dansyl method for determining the N-terminal amino acid of a peptide (see Chapter 4). This method, together with the Edman degradation, has given rise to the highly sensitive dansyl-Edman method for manual sequencing.[7] The procedure is as follows:

(i) The Edman degradation is carried out in a small glass-stoppered test tube, and is essentially as described in Section 1, but with minor modifications. The coupling reaction is carried out in 66% pyridine and the cleavage acid is trifluoracetic acid. There is no benzene wash to remove side products after the coupling reaction since small peptides would be easily extracted.

(ii) Following the cleavage step the thiazolinone derivative is extracted, (together with by-products), with n-butyl acetate *and discarded*.

(iii) The residual peptide is dried and then re-dissolved in pyridine in preparation for another cycle of the Edman degradation. However, before the PITC is added, a small aliquot ($\sim$ 5%) of the peptide solution is withdrawn and dried in a small sample tube ($2'' \times \frac{1}{4}''$). The newly liberated N-terminal amino acid is then identified in this sample by dansylation (see Chapter 4). This procedure is repeated at the start of each new cycle. An obvious disadvantage of this method is the fact that one is removing peptide at each cycle and therefore one has less material to sequence at each step. Fortunately, this is more than compensated for by the increased sensitivity of the dansyl-Edman method over the direct Edman degradation. The dansyl-Edman method is approximately one hundred times more sensitive, and consequently is the method of choice when sequencing small peptides. As little as 1 nmole of peptide can be sequenced by this method. A single cycle of the manual procedure takes approximately two and a half hours and therefore three cycles on a peptide may be conveniently achieved in a normal working day. The actual work involved in carrying out one cycle is minimal. For most of the time samples are either incubating or drying under vacuum. Since the manipulative procedures are quite straightforward, it is quite normal to sequence eight or even twelve peptides at the same time. In this way as many as 36 residues can be identified each day. The major criticism of the dansyl-Edman method is that dansyl derivatives of asparagine and glutamine are identified as the corresponding acids since they are deaminated at the acid hydrolysis stage. Residues identified as aspartic or glutamic acid during sequence analysis are, therefore, usually designated 'asx' or 'glx' since the acid and amide residues cannot be differentiated. However, Offord has shown that measurements of the electrophoretic mobility of small peptides can be used to determine the amide composition of a peptide and in many cases this additional information is sufficient to assign amide residues.[8]

The Dabsyl Method

Dabsyl isothiocyanate (4-N,N-Dimethyl aminoazobenzene 4' iso-thiocyanate, see Figure 5.3) was introduced in 1976 by Chang as an alternative to phenylisothiocyanate in the Edman degradation.[9] The thiohydantoin derivatives produced can be separated by chromatography on polyamide plates and then identified as red coloured spots by exposing the plate to HCl fumes. As little as 2 nmoles of a peptide have been sequenced by this method. This 'direct' method of sequencing therefore approaches the dansyl-Edman technique in terms of its sensitivity, and has the advantage that asparagine, glutamine and tryptophan are all determined directly by this method. One disadvantage of this reagent is that it does not react as readily with α-amino groups as does PITC. This problem has been overcome by carrying out two coupling reactions, one with dabsyl isothiocyanate and then another with PITC to ensure that all amino groups are coupled with isothiocyanate.[10] This method is not yet in routine use in many laboratories having only recently been introduced. However, it shows considerable potential and may well become a routine method in many laboratories in the future.

Figure 5.3: Dabsyl Isothiocyanate

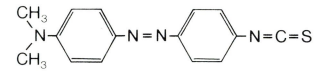

5. Automated Sequence Analysis

In an attempt to overcome the problems outlined in Section 3, and thereby increase the repetitive yield at each cycle and to increase the rate at which sequence information is generated, machines have been developed which carry out the Edman degradation automatically, and to a greater level of efficiency than can be obtained by manual methods. Two different designs of machine exist: the liquid phase (spinning cup) sequencer, and the solid phase sequencer. The design and operation of these two machines will be described separately.

The Liquid-phase (Spinning Cup) Sequencer

The first automated sequencer was described by Edman and Begg in 1967.[11] Nowadays, four spinning cup protein sequencers are commercially available, namely Jeol (Japan); Socosi (France); Illitron (USA); and Beckman (USA), but the Beckman 890 series sequencer is by far the most widely used machine having proved its success in many laboratories over a number of years. As the name suggests, all reactions take place in a spinning glass cup (Figure 5.4). The precision made cup is of Pyrex glass and has an internal diameter of 25 mm and a height of 34 mm. The cup is housed in a sealed and thermostatically maintained reaction chamber which can be connected to either a high vacuum source or a nitrogen supply. The chamber is heated by circulating hot air and a magnetic drive system is used to spin the cup (2,000 rpm). The sample to be sequenced is introduced into the spinning cup as an aqueous solution which is forced up the walls of the cup by the centrifugal force. A small undercut halfway up the cut limits the height to which the sample can spread. The cup is then placed under vacuum, when the protein sample dries as an extremely thin film on the wall of the cup. This thin film of protein is the most important feature of the spinning cup sequencer, since it provides a large surface area for reactions to occur in high yield. The film is of the order of 100 molecules thick and therefore even insoluble proteins react to a high degree of efficiency. All reagents and buffers are automatically delivered to the bottom of the spinning cup where they are forced up the sides by the centrifugal force. The time of delivery is controlled to ensure that buffers and reagents just reach the undercut (i.e. the protein sample is completely covered). The protein film can be observed in the spinning cup by the use of a stroboscope. This is particularly useful for determining whether the protein film is uniform and if the coupling buffer or cleavage acid is completely covering the sample. Following both the coupling and cleavage reactions, the protein film is dried under vacuum and then washed with the appropriate organic solvent. The solvent is delivered continuously to the base of the spinning cup where it is forced up the sides of the wall (beyond the undercut) by centrifugal force and collected by a 'scoop' (plastic tube) from a groove in the top of the cup. Benzene washes are directed to a waste bottle, whereas butyl chloride washes (containing the thiozalinone derivative) are directed to a refrigerated fraction collector. Once collected, each thiazolinone is dried down automatically in the fraction collector under a stream of nitrogen and at reduced pressure. Conversion of the dried thiazolinone to the stable PTH derivative is normally carried out by the machine operator, although an automatic device for carrying out this step has been described.[12] All solvents and reagents for use in the sequenator are highly purified chemicals which have been carefully analysed to ensure the absence of peroxides and aldehydes, and then stored under nitrogen. Commercially available sequenator chemicals are consequently very expensive to purchase. While in the machine, chemicals and reagents are stored under nitrogen and delivered to the reaction vessel under nitrogen pressure. Problems due to oxidation, Schiff base formation and insolubility are essentially overcome therefore in the sequenator. The reaction times and delivery times for solvents and reagents are all predetermined by a punched tape. Commercially available sequencers generally produce repetitive yields in the range 96–97% and require of the order of 100 to·200 nmol of peptide or protein per run. This represents about 1–2 mg of a 100 residue peptide or protein. (This figure will vary somewhat from lab to lab depending on the sensitivity of the PTH identification system available). However many workers have reported modifications to their commercially available machines all of which contribute to increased repetitive yields and reduced background in PTH samples, hence increasing sensitivity. These changes include improved vacuum systems, altered valve and delivery systems, and the use of argon in place of nitrogen.[13] Such modifications are frequently expensive and often

Figure 5.4: Design of the Sequencer Reaction Cup

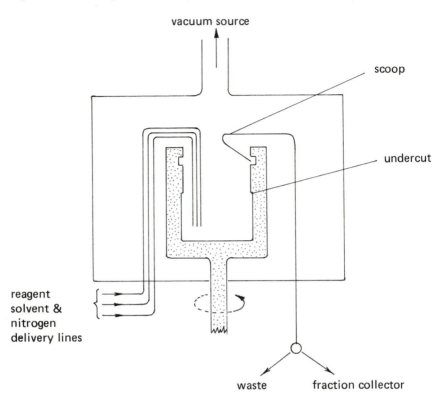

require the presence of skilled workshop facilities. However the success of such modifications cannot be denied. Sequences obtained on as little as 20 pmol of protein have been described using modified sequencers[14] and repetitive yields as high as 99% (100 residues sequenced) have been reported. However, it should be stressed that such sensitive results are produced by only a few specialist laboratories at present.

Although theoretically suitable for both peptides and proteins, difficulties have been experienced when attempting to sequence small peptides in the spinning cup sequencer. This is mainly due to the washing out of the peptide sample from the cup at the organic wash steps. A number of approaches have been taken to try to eliminate this problem.

(i) Many sequenator programmes use 'Quadrol' (a quaternary amine) as the coupling buffer. Following the coupling reaction, the buffer is removed by washing with ethyl acetate, and at this stage considerable loss of peptide can occur. This wash can be eliminated by replacing Quadrol with a volatile buffer. Two such buffers have been used successfully: N,N-dimethylbenzylamine (DMBA) and N,N-dimethylallylamine (DMAA).

(ii) It has been previously mentioned that the initial coupling reaction with PITC converts all lysine side chains to the phenylthiocarbamyl derivative. This effectively converts the hydrophilic lysine side chain to a hydrophobic side chain and consequently increases the peptide's solubility in organic solvents. This is particularly relevant for lysine rich peptides. Braunitzer has attempted to overcome this problem by carrying out the initial coupling reaction with 4-sulpho-phenylisothiocyanate (Figure 5.5). The ε-amino groups of lysine are therefore converted to a sulphophenylthiocarbamyl derivative which is hydrophilic in nature.[15] Braunitzer and co-workers have since introduced a number of improved derivatives of similar structure to modify the ε-amino groups of lysine residues.[16] Succinylation of lysine side chains (see Chapter 4) has a similar effect since it introduces a negative charge on the lysine side chains which reduces its solubility in organic solvents.

(iii) The synthetic polymer 'Polybrene' (1,5 dimethyl, 1,5-diazaundecamethylene polymethobromide) has been shown to retain small peptides in the spinning cup.[17] As little as 3 mg of this compound is added directly to the peptide solution and has proved remarkably successful in preventing peptide losses. This method seems to be the best available to date for preventing wash-out from the sequenator cup.

Figure 5.5: Braunitzer Reagent (4-sulpho-phenylisothiocyanate)

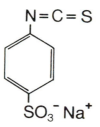

The cost of a protein sequencer in 1982 is approximately £50,000. This, together with the high cost of sequenator chemicals, makes it an expensive technique. Taking into account the cost of the technical assistance many laboratories quote a figure in excess of £10 per residue as the cost of automated sequencing. However, the sequencer does have the tremendous advantages that long sequences can be determined. The strategy for determining the sequence of a protein when a sequencer is available is therefore to prepare a small number of large peptides from digests of the native protein. This makes one's task generally much easier than if only manual sequencing facilities are available, since in this case many hundreds of small peptides have to be isolated from a number of different digests before sufficient overlaps can be obtained to give the final sequence. Such a procedure is invariably routine, and time consuming.

The Solid-phase Protein Sequencer

The solid phase sequencer was designed essentially for the automatic sequencing of peptides that contain about 30 residues or less; i.e. those that tend to be 'washed-out' in the spinning cup sequenator. However, some proteins have been sequenced on the solid phase machine. The first solid phase sequencer was described by Laursen in 1971[18] and there are now three commercially available versions of this instrument: Sequemat (USA); L.K.B. (UK) and Socosi (France). In the solid phase sequencer the peptide is covalently bound to an inert support and packed in a micro-bore column (~ 2.0 mm internal diameter × 150 mm). The attached peptide is then subjected to repeated cycles of the Edman degradation by pumping the relevant chemicals and solvents through the column. The column is maintained at the required temperature by circulating water through a surrounding glass jacket. For the coupling step the PITC and buffer solutions are pumped from separate reservoirs, mixed and then passed through the column. When coupling is complete, excess reagents, together with any by-products, are removed by pumping solvents, normally methanol and dichloroethane, through the resin. The released thiazolinone is washed from the column with methanol, collected in a fraction collector, and converted manually to the PTH derivative. 'Washing out' of the peptide is of course impossible since it is chemiclly bound to the support resin. One disadvantage of solid phase sequencing over 'spinning cup' sequencing is that in solid phase sequencing one has the extra step of chemically modifying the peptide (i.e. attaching it to the support resin). As yet the methods available for attaching peptides to resins are not completely satisfactory in all cases and further development in this area will be advantageous. Problems relate particularly to the yield of peptide attachment to the support. This is frequently below 100%, sometimes as low as 20–30%. Since the rate limiting step in any sequencing is the production of peptides in sufficient quantity for sequencing, to risk losing a large percentage of one's purified peptide at this step is unacceptable to many workers. Two different types of column support can be used. These are polystyrene and porous glass beads. Four of the more popular support mediums in use are shown in Table 5.2, two being polystyrene derivatives and two glass bead derivatives. It should be pointed out, however, that new support resins and methods of attachment are continually being developed and new methods appear frequently in the literature. Only the passage of time will tell which of the methods are most generally useful. Attachment to the resin can be made either via a C-terminal residue or through a side chain group of the peptide. Three examples of these types of attachment are described here:

Attachment through C-terminal Homoserine Lactone. Peptides obtained by cleaving proteins with cyanogen bromide terminate with homoserine (with the exception of the peptide derived from the C-terminus of the protein). Treatment of these peptides with trifluoracetic acid converts the C-terminal

Table 5.2: Some Supports commonly used in Solid Phase Sequencing

Support	Structure

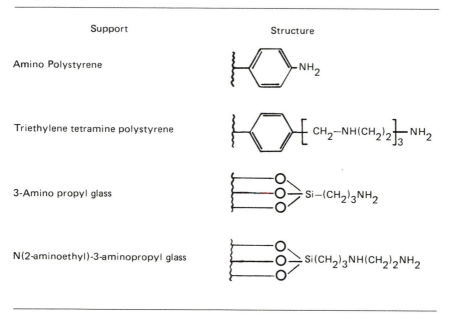

Support	
Amino Polystyrene	
Triethylene tetramine polystyrene	
3-Amino propyl glass	
N(2-aminoethyl)-3-aminopropyl glass	

homoserine to homoserine lactone. This lactone can now be coupled directly to a polymeric amine, e.g. triethylene tetramine polystyrene (Figure 5.6), by simply mixing peptide and resin together in dimethylformamide.[19] Following coupling, excess amino sites on the resin are blocked with methyl isothiocyanate and the peptide is ready for sequencing. This is the simplest of all the coupling reactions and occurs in high yield. However, the fact that it can only be applied to cyanogen bromide peptides does limit its usefulness.

Attachment through the C-terminal Carboxyl Group. This attachment is usually achieved by using a water-soluble carbodiimide, e.g. N-ethyl-N' (3-dimethyl aminopropyl) carbodiimide[20] (Figure 5.7). Protein terminal amino groups are first blocked with a reversible blocking group. Treatment of the peptide with carbodiimide results in reaction with both the C-terminal carboxyl groups of the peptide and the side chain carboxyl groups of aspartic and glutamic acid residues. The C-terminal carboxyl group forms an oxozolinone which will react with amino groups of a resin, whereas the side chain carboxyl groups give an O-acyl urea derivative which subsequently rearranges to give an inactive N-acyl urea which will not react with resin amino groups. (This is fortunate since, if the side chain carboxyl groups did react with the resin, these residues would not be identified in the sequencing procedure). In practice, a number of side reactions can also occur and, unfortunately, the coupling efficiency is variable.

Attachment via Lysine Side Chain Groups. In this method, peptides are coupled using p-phenylenediisothiocyanate to crosslink resin and peptide amino groups[21] (Figure 5.8). The peptide attaches to the resin via both its α-amino group and the ε-amino side chains of the lysine residues. Treatment with anhydrous trifluoracetic acid cleaves the N-terminal amino acid from the chain leaving the peptide with a free α-amino group, but still attached to the resin via the ε-amino side chains. The N-terminal residue remains

Figure 5.6: Attachment of Cyanogen Bromide derived Peptides to an Inert Support

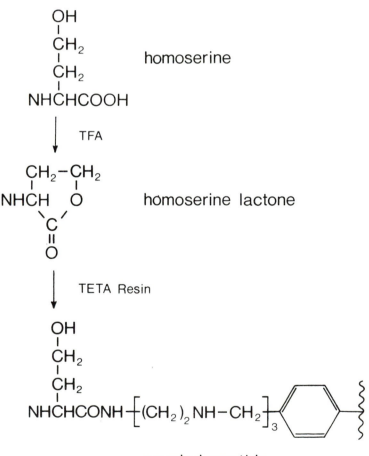

coupled peptide

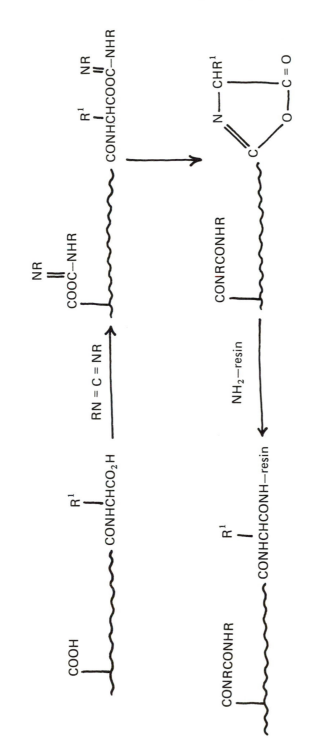

Figure 5.7: The Coupling of Peptides to an Insoluble Support using a Carbodiimide

Figure 5.8: The Attachment of Peptides to Solid Supports via Lysine Side Chains using Phenylenediisothiocyanate

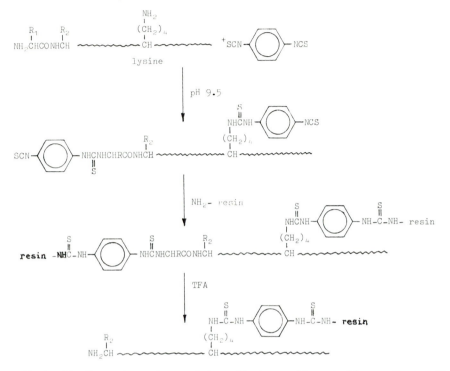

attached to the resin as do any internal lysine residues and hence these will appear as gaps in the sequence. This method is therefore of little use for peptides containing large numbers of lysine residues. However, it often gives high yield, up to 100%, and is particularly useful for tryptic peptides which end in a lysine residue. Peptides not containing lysine obviously cannot be used for this method.

Compared to the spinning cup sequencer, the solid phase machine is cheaper to purchase (about a quarter of the price) and cheaper to run, since it requires smaller volumes of solvents and reagents. However, despite these advantages, solid phase sequencing has not yet achieved the popularity of the spinning cup machine. This is mainly due to the problems associated with coupling peptides to the resin supports. In practice sequence lengths of 25–30 residues can be obtained by solid phase sequencing using 50–100 nmol of peptide. As yet, therefore, solid phase sequencing cannot produce such lengthy sequences as, or rival the sensitivity of, the modified commercial spinning cup machines, although sensitivity is comparable to that of the standard commercial machines. However, it must be stressed that the solid phase sequencer has not been in use as long as the spinning cup sequencer and that further developments of this technique are to be expected in the future.

The Gas-liquid Solid-phase Peptide and Protein Sequencer

With the ever increasing sophistication and sensitivity of techniques in molecular biology there is a continuing demand for greater sensitivity in protein sequence analysis (e.g. to the point where extensive sequence data can be obtained on protein bands eluted from acrylamide gels). The fact that chemists have taken up the challenge to increase sensitivity is shown by the recent publication of a gas-liquid solid phase sequencer.[22] Gas phase reagents are used at both the coupling and cleavage steps of the Edman degradation and the sample to be sequenced is embedded in a matrix of Polybrene dried onto a porous glass fibre disc located in a small cartridge style reaction cell. The protein or peptide, although not covalently attached to the support, is essentially immobile throughout the degradative cycle since only relative apolar, liquid phase solvents pass through the cell. This instrument has been used to provide sequence data on as little as five picomoles of protein and can perform extended sequence runs (up to 55 residues) on subnanomolar quantities of protein. Sequence data (> 30 residues) has also been obtained on subnanomolar quantities of protein electrophoretically eluted from Coomassie blue stained SDS poly-acrylamide gels. The cycle time is about 50 minutes with a repetitive yield of 98%, and the authors believe that sequencing sensitivity can be increased to the one pmol level. Such an instrument is obviously another important step in the continual improvement in sensitivity of automated sequence analysers. However, it should be stressed that these machines, like the modified commercially available spinning cup sequencers are at present only to be found in a small number of specialised laboratories devoted to protein sequence analysis. These machines are not yet, or indeed are unlikely ever to be, commercially available.

6. Blocked N-terminal Groups

A number of proteins and peptides have been shown to have their N-terminal amino groups blocked. Such molecules are obviously unreactive in the Edman degradation, being unable to couple with phenylisothiocyanate. Neither can these N-terminal amino acids be identified by the dansyl method. Among the more common blocking groups are acetyl (e.g. rabbit muscle myosin) and formyl (e.g. melittin) groups, and pyroglutamic acid.

Acetyl Groups

Since there is no chemical method for selectively removing blocked acetyl groups, N-terminal sequence analysis of the native protein is not possible. N-terminal sequences are therefore determined by proteolytic degradation of the native protein followed by isolation of the blocked N-terminal peptide. If this peptide does not contain a lysine residue it will not be ninhydrin positive and alternative peptide stains such as the starch/KI stain should be used for its identification. When isolated, the blocked peptide should be further degraded, if possible, with other enzymes to give a di- or tri-peptide. This may then be sequenced from the C-terminus (see Section 8) or, if facilities exist, by mass spectrometry. Mass spectrometry has the advantage of identifying the nature of the blocked N-terminal amino acid, whereas C-terminal sequencing leaves the blocked group unidentified. If mass spectrometry facilities are not available, acid hydrolysis of the N-terminal peptide will release acetic acid which can be extracted with ether and identified, and the amino acid composition of the hydrolysate will identify which amino acid is blocked, since the other aminoacids in the sequence will have been previously determined by C-terminal sequencing.

Formyl Groups

N-terminal formyl groups pose less of a problem than acetyl groups since they may be removed by mild acid hydrolysis under conditions which do not cause serious cleavage of peptide bonds within the protein. Alternatively, N-terminal formyl groups may be removed by a deformylase enzyme which can be isolated from *E. coli*.

Figure 5.9: The Cyclisation of N-terminal Glutamine to Pyroglutamic (Pyrrolidone Carboxylic) Acid

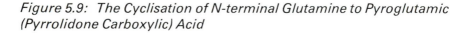
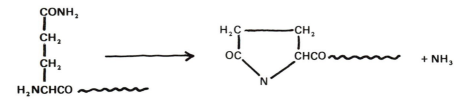

Pyroglutamyl Derivatives

Under certain conditions, glutamine (either as the free amino acid or when the N-terminal amino acid of a peptide or protein) will undergo cyclisation to the pyroglutamyl derivative (Figure 5.9). This cyclised derivative does not have a free amino group and therefore is not amenable to sequence determination. Unfortunately, partial cyclisation sometimes occurs during sequencing procedures when glutamine is the newly liberated N-terminal amino acid. That cyclisation has occurred is apparent when the yield of PTH derivatives drops dramatically at a glutamine residue. Unfortunately, nothing can be done when this occurs during the sequence procedure. A number of naturally occurring peptides and proteins have been reported to have an N-terminal pyroglutamyl residue, although it is always difficult to prove that cyclisation of N-terminal glutamine has not occurred during the isolation of the protein. N-terminal sequence analysis can only be carried out on such proteins if the pyroglutamyl residue is removed. This can be done using the enzyme pyrrolidonyl peptidase (from Pseudomonas fluorescence) which specifically hydrolyses the N-terminal pyroglutamyl linkage.

7. Sequence Determination by Mass Spectrometry

A relatively recent innovation in the field of protein sequence determination has been the introduction of mass spectrometry as a sequencing tool. Mass spectrometry can cope with peptides in the range two to 15 amino acids in length, and occasionally longer.[23] The peptide is initially chemically modified by permethylation using 'dimsyl' sodium (methyl sulphinyl sodium) and sodium iodide. Following reaction, the permethylated derivative is extracted into chloroform and its spectrum recorded in this solvent. Permethylation of the peptide has the effect of increasing volatility of the peptide (an essential requirement for mass spectrometry) due to the inhibition of interchain hydrogen bonding, and of stabilising the fragmentation pathway that occurs in the mass spectrometer. Fragmentation of the modified peptide occurs essentially at peptide bonds giving rise to 'sequence ions' which may be identified. However, it would be foolish to try to oversimplify the interpretation of mass spectrometer data. Many other fragmentation pathways occur and the exact nuances of interpretation are too detailed for discussion here. Suffice to say that considerable skill and experience are necessary for the interpretation of every signal in the spectrum. For this reason in particular, sequencing by mass spectrometry is not, as yet, in common use. However, new developments in computer automation of the interpretation of mass spectra data should help obviate this problem.

As well as producing peptide sequences (the protein dihydrofolate reductase, M.W. 18,000, was completely sequenced from mass spectrometry data), mass spectrometry can also provide additional information over that provided by traditional sequencing techniques. For example, mass spectrometry can provide information on both peptide and carbohydrate structure. Glycopeptides may, therefore, be analysed in a single experiment. γ-carboxyglutamic acid (a constituent of prothrombin) was first identified by mass spectrometry.[24] This amino acid is easily decarboxylated to glutamic acid and this had indeed occurred during conventional sequence analysis of prothrombin where γ-carboxyglutamic acid residues were identified as glutamic acid residues. The occurrence of γ-carboxylglutamic acid in prothrombin was unknown prior to analysis by mass spectrometry. The sensitivity of sequence analysis by mass spectrometry is superior to that of conventional sequencing techniques;

5–30 nmol of a peptide are usually required and in favourable cases picogram quantities of material have been sequenced. However, the major attraction of sequencing by mass spectrometry is the fact that *mixtures* of peptides can be sequenced. This is achieved by fractional distillation of the peptide mixture on the probe tip within the mass spectrometer. During this distillation the signals associated with any one peptide in the mass spectrum rise and fall together at different rates when compared to other components in the mixture. Each component can therefore be identified separately. This method is best suited to mixtures of up to five peptides. Since the production of purified peptides is always the rate-limiting step in any sequence analysis, mass spectrometry can greatly increase the rate at which sequence data is obtained. Proteolytic digestion of a native protein, followed by a single separation step (such as a single ion-exchange column step) can provide fractions immediately suitable for analysis by mass spectrometry. In this way the need for the laborious separation of purified peptides is obviated. It has been estimated that mass spectrometry may reduce the time taken to sequence a protein of M.W. 20,000 to two months. It should be apparent from the above that mass spectrometry is obviously a technique with enormous potential for the future.

8. C-terminal Sequence Analysis

Unlike N-terminal sequencing, there is no effective chemical method for selectively removing amino acids sequentially from the C-terminus of a peptide or protein. However, methods do exist for determining either the C-terminal amino acid or a limited sequence at the C-terminus or a peptide or protein. These are described below.

Figure 5.10: Determination of the C-terminal Residue of a Polypeptide by Hydrazinolysis

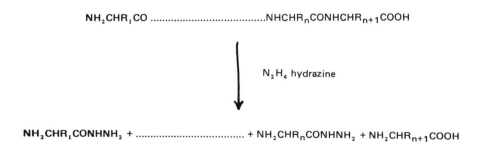

Hydrazinolysis

This method was first introduced by Akabori[25] and provides information on the nature of the C-terminal amino acid of peptides and proteins (Figure 5.10). The peptide or protein is treated with anhydrous hydrazine at 100°C for twelve hours. During this time the hydrazine converts all the amino acid residues, except the C-terminal one, into their respective hydrazides. Following the reaction period, excess hydrazine is removed under vacuum, the residue dissolved in water, and amino acid hydrazides removed by extraction with benzaldehyde. The amino acid remaining in the aqueous phase is then identified by amino acid analysis. This method has proved successful on many occasions, but unfortunately is not applicable to C-terminal arginine, glutamine and asparagine.

Tritium Labelling

This method was introduced by Matsuo[26] and provides information on the nature of the C-terminal amino acid (Figure 5.11). Treatment of the peptide or protein with acetic anhydride results in the formation of a C-terminal oxazolone. Treatment of the oxazolone with pyridine, in the presence of T_2O, causes base catalysed racemisation with concomitant ring opening, leaving tritium incorporated within the C-terminal amino acid. The peptide or protein is then hydrolysed into its constituent amino acids which are separated by amino acid analysis, and the radioactive amino acid identified.

The Use of Carboxypeptidases

Whereas the previous two methods only identify the C-terminal amino acid, the use of carboxypeptidase enzymes can provide information on the *sequence* at the C-terminus of a peptide or protein. The carboxypeptidases are a group of exopeptidases which can remove amino acids sequentially from the C-terminus of peptides and proteins. Four carboxypeptidases (carboxypeptidase, A, B, C and Y) have been identified and differ with respect to their specificity and source.[27] Carboxypeptidase A (pancreatic) removes all C-terminal amino acids excpet arginine, lysine and proline, whereas carboxypeptidase B (pancreatic) removes arginine and lysine rapidly with little effect on other amino acids. Carboxypeptidases A and B are commercially available and treatment of polypeptides with a mixture of these enzymes will release all C-terminal amino acids except proline. Carboxypeptidase C (from citrus fruits) and carboxypeptidase Y (baker's yeast) release all C-terminal amino acids but only carboxypeptidase C is generally commercially available. Experimentally, the peptide or protein is treated with an appropriate mixture of carboxypeptidase enzymes and aliquots removed at regular time intervals and the reaction stopped. Each sample is quantitatively analysed for the presence of free amino acids by amino acid analysis and a graph of amino acids liberated against time plotted. From the relative rate at which individual amino acids are liberated the sequence of amino acids at the C-terminus may be deduced.

Unfortunately, because any given enzyme releases different C-terminal amino acids at different rates, asynchrony is soon introduced into the relative rate at which amino acids are liberated and therefore a C-terminus sequence of three or four residues is usually the best that can be obtained. However, such information is often extremely useful.

Figure 5.11: Tritium Labelling Method for C-terminal Amino Acids

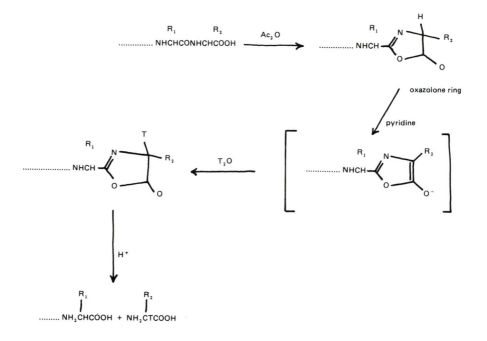

Further Reading

1. S.B. Needleman (ed.), *Protein Sequence Determination* (Springer-Verlag, 1970).
2. S.B. Needleman (ed.), *Advanced methods in protein sequence determination*, (Springer-Verlag, 1977).
3. L.R. Croft, *Introduction to Protein Sequence analysis*, (John Wiley & Sons, 1980).
4. C. Birr (ed.), *Methods in peptide & protein sequence analysis*, (Elsevier/North Holland, 1980).
5. C.H.W. Hirs (ed.), *Enzyme Structure*, Methods in Enzymology, 11, (Academic Press, 1967).
6. C.H.W. Hirs (ed.), *Enzyme Structure Part B*, Methods in Enzymology, 25, (Academic Press, 1972).
7. Z. Deyl, 'Sequence Analysis of Proteins & Peptides', *J. Chromatog.*, *127* (1976), 91–132.

References

1. P. Edman, 'A Method for the Determination of the Amino Acid Sequence in Peptides', *Acta Chem. Scand.*, *4* (1950), 283–293.
2. C.L. Zimmerman, E. Appella and J.J. Pisano, 'Rapid Analysis of Amino Acid Phenylthiohydantoins by High Pressure Liquid Chromatography', *Anal. Biochem.*, *77* (1977), 569–573.
3. J. Rosmus and Z.J. Deyl, 'Chromatography of N-terminal Amino Acids and Derivatives', *J. Chromatog.*, *70* (1972), 221–339.
4. G. Rosean and P. Pantel, 'Revelation coloree des spots de phenylthiohydantoin d'acides amines', *J. Chromatog.*, *44* (1969), 392–395.
5. K.D. Kulbe, 'Micropolyamide thin-layer chromatography of phenylthiohydantoin amino acids (PTH) at sub microgramme levels', *Anal. Biochem.*, *20* (1974), 89–102.
6. O. Smithies, D. Gilson, E.M. Fanning, R.M. Goodfleish, J.G. Gilman and D.L. Ballantyne, 'Quantitative Procedures for use with the Edman-Begg sequenator', *Biochemistry*, *10* (1971), 4912–4921.
7. B.S. Hartley, 'Strategy and Tactics in Protein Chemistry', *Biochem. J.*, *119* (1970), 805–822.
8. R.E. Offord, 'Electrophoretic Mobilities of Peptides on Paper and their use in the determination of amide groups', *Nature*, (1966), 591–593.
9. J.Y. Chang, E.M. Creaser and K.W. Bentley, '4,N,N-Dimethyl-aminoazobenzene 4'-isothiocyanate, A new chromophoric Reagent for Protein sequence analysis', *Biochem. J.*, *153* (1976), 607–611.
10. J.Y. Chang, D. Brauer and B. Wittmann-Liebold, 'Micro-sequence analysis of peptides and proteins using 4-NN-Dimethyl-aminoazobenzene 4'-isothiocyanate/phenylisothiocyanate double coupling methods', *FEBS Letts*, *93* (1978), 205–214.
11. P. Edman and G. Begg, 'A protein sequenator', *Eur. J. Biochem.*, *1* (1967), 80–91.
12. B. Wittmann-Liebold, H. Graffunder and H. Kohls, 'A device coupled to a modified sequenator for the automated conversion of Anilinothiazolinones into PTH amino acids', *Anal. Biochem.*, *75* (1976), 621–633.
13. M.W. Hunkapiller and L.E. Hood, 'Direct Microsequence Analysis of Polypeptides using an improved Sequenator', *Biochemistry*, *17* (1978), 2124–2133.
14. M.W. Hunkapiller and L.E. Hood, 'New Protein Sequenator with Increased Sensitivity', *Science*, *207* (1980), 523–525.
15. G. Braunitzer, B. Schrank and A. Rhufus, 'On the complete automatic sequence analysis of peptides using quadrol', *Hoppe-Seyler's Z. Physiol. Chem.*, *352* (1971), 1730–1732.
16. G. Braunitzer, B. Schrank, S. Petersen and U. Peterson, 'Automatic Sequencing of Insulin', *Hoppe-Seyler's Z. Physiol. Chem.*, *354* (1973), 1563–1566.
17. G.E. Tarr, J.F. Beecher, M. Bell and D.J. McKean, 'Polyquaternary amines prevent peptide loss from sequenators', *Anal. Biochem.*, *84* (1978), 622–627.
18. R.A. Laursen, 'Solid-phase Edman degradation: an automatic peptide sequencer', *Eur. J. Biochem.*, *20* (1971), 89–102.
19. M.J. Horn and R.A. Laursen, 'Attachment of carboxyl-terminal homoserine peptides to an insoluble resin', *FEBS Letts*, *36* (1973), 285–288.

20. A. Previero, M.A. Coletti-Proviero and R.A. Laursen, 'Solid-phase sequential analysis: specific linking of acidic peptides by their carboxyl ends to insoluble resins', *FEBS Letts, 33* (1973), 135–138.
21. R.A. Laursen, M.J. Horn and A.G. Bonner, 'The use of p-phenyl diisothiocyanate to attach lysine and arginine-containing peptides to insoluble resins', *FEBS Letts, 21* (1972), 67–71.
22. R.M. Hewick, M.W. Hunkapiller, L.E. Hood and W.J. Dreyer, 'A gas-liquid solid phase Peptide and Protein Sequenator', *J. Biol. Chem., 256* (1982), 7990–7997.
23. H.R. Morris, 'Biomolecular structure determination by mass spectrometry', *Nature, 286* (1980), 447–452.
24. S. Magnusson, L. Sottrup-Jensen, T.E. Peterson, H.R. Morris and A. Dell, 'Primary structure of the vitamin K-dependent part of prothrombin', *FEBS Letts, 44* (1974), 189–193.
25. S. Akabori, K. Ohno and K. Narita, 'On the hydrazinolysis of proteins and peptides', *Bull. Chem. Soc. Japan, 25* (1952), 214–218.
26. J. Matsuo, Y. Fujimoto and T. Tatsuno, 'A novel method for the determination of C-terminal amino acids', *Biochem. Biophys. Res. Commun., 67* (1966), 404–412.
27. R.P. Ambler, 'Enzymatic Hydrolysis with Carboxypeptidases', in *Methods in Enzymology* (Academic Press, New York), 25, 262–71.

6 THE EXTRACTION AND FRACTIONATION OF RNA

Robert J. Slater

CONTENTS

1. General Introduction

Biological materials contain two forms of nucleic acid, namely DNA and RNA. It has been known for a number of years that RNA molecules are involved in the expression of the genetic material (DNA) by directing protein synthesis. Knowledge of the metabolism and mode of action of RNA is therefore crucial to our understanding of cell division, growth and development, and the future progress of genetic engineering.

There are many different types of RNA present in the cell and they are generally classified according to function or by size (molecular weight or sedimentation coefficient, S). The three main types of RNA according to function are ribosomal, transfer and messenger RNA (rRNA, tRNA and mRNA respectively). Ribosomal RNA molecules are structural components of ribosomes, and constitute the bulk of cellular RNA (approximately 75%). Each ribosome contains three different molecules of rRNA classified by size: 28S (or 25S) 18S and 5S in eukaryotic, cytoplasmic ribosomes or 23S, 16S and 5S in prokaryotic-type ribosomes. Transfer RNAs are small molecular weight molecules (approximately 25,000 Daltons or 4S) responsible for the supply of specific amino acids to the ribosomes. Messenger RNA molecules are of heterogenous molecular weight, are generally present as a small percentage (one to five per cent) of cellular nucleic acid and are responsible for directing assembly of the correct amino acid sequence during protein synthesis.

There has been considerable research interest in the synthesis and function of RNA in recent years. An important aspect of this research concerns the extraction and fractionation of RNA and numerous methods can be found in the research literature. In the majority of cases, the methodology described is suited to a particular organism or tissue and RNA type. There are, however, a number of general principles common to all the methods given. This chapter is intended to introduce the reader to these general principles and will point the way to more specific methodology in the references. The following sections will outline the methods involved in the preparation of RNA, the special considerations pertinent to the extraction of mRNA, the estimation of mRNA content and the fractionation of RNA.

2. The Extraction of Total RNA

Introduction

The prime objective in the extraction of RNA is the quantitative recovery of pure RNA in an undegraded form. The principal obstacle in the way of achieving this objective is the fact that the majority of RNA molecules in the cell exist as RNA-protein complexes (often called ribonucleoprotein or RNP) such as ribosomes and polyribosomes. The interactions between nucleic acids and protein are complex and not fully understood but they are sufficiently strong to require rather severe chemical conditions for their destruction. The extent to which the required RNA can be separated from protein molecules is one very important factor contributing to the overall success of the method used. Three basic methods are commonly used to deproteinise RNP or whole cells. These employ: phenol as a protein denaturant; a highly purified protease solution to degrade protein from the RNP complex or strong salts to dissolve the nucleic acid and precipitate denatured protein simultaneously.

All three approaches are very simple practically and require little or no specialised equipment. The success of the methods relies not simply on the successful removal of protein but also on avoiding ribonuclease degradation of the RNA during extraction, purification and storage of the final product.

The problem of ribonuclease activity during nucleic acid extraction is sufficiently important to warrant some discussion. Ribonuclease is present in most living cells and often in the extracellular fluid, such as in cell cultures or perspiration. It is of the utmost importance, therefore, that the enzyme activity is inhibited as rapidly and effectively as possible. It is for this reason that many RNA extraction procedures include the use of detergents, such as sodium dodecylsulphate (SDS), that are capable of denaturing the ribonuclease. It must be remembered however that ribonuclease can renature and become active again following removal of the detergent. A number of basic precautions are necessary to avoid RNA degradation through ribonuclease activity:

(i) RNA is extracted and manipulated under conditions of the greatest cleanliness (wearing disposable rubber gloves is a useful precaution against ribonuclease in the perspiration of fingers)

(ii) Solutions are autoclaved where possible, particularly those not containing detergents or deproteinising agents.

(iii) Glassware is pretreated by autoclaving, baking in an oven (150°C, 30 minutes) or soaking in a solution of 0.5% diethylpyrocarbonate (a powerful protein denaturant) followed by incubation at 100°C for five minutes.

(iv) RNA is stored under conditions that minimise ribonuclease activity, for example in the presence of SDS or as a precipitate under ethanol at − 18°C.

(v) Chemicals of the highest purity are used to keep the possibility of ribonuclease contamination to a minimum.

It is particularly important to observe these precautions when working in a laboratory that is involved in the purification of DNA because ribonuclease treatment is a commonly used procedure, thereby contaminating glassware, pipettes, etc.

From the point of view of starting material for RNA extraction there are two basic alternatives. Total nucleic acids can be extracted from the tissue, followed by removal of DNA and purification of the RNA under study. This is the usual method for the isolation of large quantities of rRNA or total mRNA. The alternative procedure is to begin the isolation with subcellular fractionation. This approach is commonly used for the isolation of specific mRNA or for isolation of RNA from specific organelles (e.g. chloroplasts and mitochondria). Transfer RNA can be isolated successfully by either approach but the fact that tRNA occurs in a free form means that the bulk of the cellular material can be removed by centrifugation as a first step in tRNA isolation.[1]

Phenol methods

Phenol methods are particularly useful for the extraction of total nucleic acids or total RNA. The basic method described here is applicable to bacteria, fungi, plants and animal tissues. Homogenisation can be carried out by relatively gentle procedures such as mortar and pestle or Potter homogeniser if an effective detergent lysing medium is used. The lysing solution described by Kirby[2] is still commonly used today. This solution contains a strong detergent (sodium tri-isopropylnaphthalene sulphonate 1%) for cell lysis and inhibition of ribonuclease, a chelating agent (sodium 4-aminosalicylate, 6%) for disruption of nucleoprotein and sodium chloride 81%) to inhibit denaturation of nucleic acid. A small amount of phenol mixture (5%), described later, is included to increase the solubility of the detergent. This solution will readily lyse gram-negative bacteria but is best preceded by lysozyme treatment for gram-positive organisms. A tissue to volume ratio of approx. 1:10 is convenient and is very effective at inhibiting ribonuclease. If higher tissue to volume ratios are essential to the experiment then the use of lysing medium at double strength is recommended. Homogenisation at low temperature (0–4°C) assists in preventing ribonuclease activity. Successful lysis will be accompanied by an increase in the viscosity of the medium due to release of DNA.

Following homogenisation the cell lysate is deproteinised by shaking with an equal volume of redistilled phenol or phenol mixture. The latter solution contains 500 g phenol crystals (should be colourless) 70 ml redistilled m-cresol to act as an antifreeze, 0.5 g 8-hydroxyquinolene to inhibit phenol oxidation and is saturated with water (150 ml) to prevent reduction in volume of the aqueous layer during subsequent phase partition.

Deproteinisation is most effectively carried out at room temperature but the phases are best separated by centrifugation at 0–4°C.

Following centrifugation, three distinct phases should be visible: an upper aqueous phase that contains nucleic acids, a lower phenol phase that contains hydrophobic cell components and a middle layer of denatured protein. The aqueous layer is removed and deproteinised again by repeating the above procedure with another volume of phenol mixture.

If a quantitative recovery of nucleic acid is required the contents of the first centrifuge tube, containing the denatured protein and phenol layer, can be re-extracted by shaking with lysing medium and the aqueous layer combined with the previous extract and deproteinised again as before.

The aqueous phase produced by this procedure contains, in addition to the water-soluble, lysing medium components: total cellular nucleic acid, small molecular weight, soluble cellular components and some polysaccharides (in particular glycogen if liver from an unstarved animal was the starting material). Addition of two volumes of absolute ethanol, followed by precipitation of nucleic acid at −18°C overnight will remove the majority of the contaminants. Traces of phenol can then be removed by washing the precipitate in 70% alcohol containing 0.5% SDS or by re-precipitation from 0.15M sodium acetate buffer (pH 6.0) with two volumes of alcohol as before. Ethanol-insoluble carbohydrate is generally removed during subsequent fractionation of nucleic acids but differential solubility methods can be used.[3] DNA can be removed by dissolving the nucleic acid precipitates in a neutral buffer containing magnesium ions and incubating with a commercially available, highly purified deoxyribonuclease preparation. The enzyme is active at low temperature so the incubation can be done at 0–4°C to reduce the chance of RNA degradation through ribonuclease activity. The deoxyribonuclease is removed by phenol extraction as previously described.

The above methodology has been described in some detail in order to illustrate the function of the various chemical components and steps in the procedure. The general principles apply to the extraction of all forms of nucleic acid. There are, however, many modifications to the basic method described for the preparation of particular forms of RNA. For example, magnesium ions are included in the lysing medium if rRNA is required from chloroplasts[4] and significant modifications are required for the isolation of mRNA (see Section 3 'Phenol Methods'). Less drastic conditions can be used if the nucleic acids of eukaryotic nuclei are not required.[5] In this case an homogenising medium containing the detergent sodium dodecyl sulphate (0.5–2.0%) without a chelating agent can be successfully used and is a common procedure for the isolation of RNA from ribosomes. Protein can then be removed by phenol extraction as above or by purifying the RNA through a sucrose gradient (see Section 5, D).

The detergent/phenol extraction procedure is a good general method with widespread application. There are, however, a number of disadvantages with the technique. In many cases traces of protein, particularly basic proteins, are extracted with the nucleic acid. This can be a serious problem if the RNA is to be used as a template for *in vitro* translation, cDNA synthesis or as a hybridisation probe. In these cases contaminating protein can be removed by additional procedures such as RNA fractionation or, for rRNA, cellulose nitrate filtration.[6] Other disadvantages are that phenol extraction tends to result in the aggregation of RNA (see Section 5) and poly (A)-containing RNA can be resistant to phenol extraction (see Section 4). For these reasons, alternative methods for the extraction of RNA are often employed.

Proteinase K Methods

One alternative to the basic phenol method for extraction of RNA is to use proteinase K for the deproteinisation of cell homogenates, organelles or RNP.[7] Proteinase K is a very active non-specific proteolytic enzyme extracted from the culture medium of the fungus *Tritirachium album limber*. The enzyme is commercially available, contains very little detectable ribonuclease activity and is active under a wide range of conditions. Metal cations or EDTA have little or no effect on proteolytic activity and maximum activity can be recorded over a broad pH range (7.4–11.5). The most significant properties from the point of view of RNA extraction are the effect of temperature and SDS. The enzyme is highly active up to 65°C and is not inhibited by concentrations of SDS (0.1%) that are sufficiently high to inactivate ribonuclease and deoxyribonuclease. In fact, SDS can increase the activity of the enzyme, probably by the denaturing effect on the substrates.[8] The combination of SDS and proteinase K is particularly powerful at inactivating ribonuclease and the method is therefore very useful for the preparation of high molecular weight RNA. The cell homogenate or sub-cellular fraction is incubated in an appropriate buffer containing proteinase K (200 μg/ml) and SDS (0.1%). The incubation time (30 minutes to overnight) and temperature (25°C–65°C) are dependent on the system concerned and the protein content of the starting material. Any protein remaining at the end of the incubation is removed by exploiting the differential solubility of nucleic acids and proteinase K in lithium chloride.[9] Ribosomal and messenger RNA are precipitated by this salt, whereas proteinase K and tRNA remain in solution. For this reason proteinase K methods are less applicable to the isolation of tRNA.

Strong Salts

A third technique for the isolation of RNA will briefly be described here. This method involves the use of strong salts and is particularly suited to relatively small-scale preparations such as the extraction of RNA from ribosomes or viruses. The technique is based on the fact that RNA, unlike protein, is soluble in high concentrations (4M) of guanidinium chloride.[10] The cell homogenate is suspended in four volumes of a 5M solution of the guanidinium salt with the optional inclusion of a detergent such as N-lauryl sarcosine (4%) to assist disruption of the ribonucleoprotein. Protein is removed by centrifugation and the RNA is recovered from the supernatant by ethanol precipitation. For small-scale preparations the RNA solution can be directly purified by centrifugation overnight at 125,000 g through a 5.7M CsCl cushion containing 100mM EDTA.

A similar technique for the direct isolation of RNA for small tissue samples has been described by Glisin.[11] In this procedure an SDS homogenate of the tissue is centrifuged over a cushion of CsCl to obtain a pellet of RNA. The principal disadvantage of the method is that aggregation of RNA can occur.

Yield, Purity and Storage of RNA

The yield and purity of an RNA preparation can be assessed by monitoring the absorbance of ultra-violet light by a solution of the nucleic acid. A pure RNA solution should give a 260 nm : 280 nm ratio of two; an absorbance reading at 260 nm of 1.0 is equivalent to approximately 45 μg/ml.

RNA is best stored as a precipitate under 70% alcohol containing 0.15M sodium acetate. These conditions maintain secondary structure and prevent ribonuclease degradation.

3. The Extraction of mRNA

Introduction

The extraction of mRNA involves many of the basic principles of extraction of total RNA. There are, however, a number of special considerations that need to be taken into account. First, except in very specialised cells, mRNA consists of many thousands of different molecules with respect to base sequence and size. Secondly, mRNA can be particularly susceptible to RNase attack, and, thirdly, the existence of poly(A) tracts at the 3' end of the molecule affect the chemistry of the RNA and therefore the chemistry of the extraction procedure. For these reasons the isolation and purification of specific mRNA has to date been most successful from specialised cells containing large quantities of the required mRNA. One of the first examples of this approach was the isolation of globin mRNA from reticulocytes.[12] In all preparations of mRNA, however, an early step is required to separate mRNA from the bulk of cellular nucleic acids (DNA, rRNA and tRNA). This can be achieved by selective extraction procedures, cell fractionation, affinity chromatography or RNA size fractionation. In many cases the particular mRNA under study is separated from the total mRNA on the basis of size by sucrose gradient centrifugation or preparative gel electrophoresis. The following sections describe the common approaches to isolating mRNA.

Phenol Methods

It has been observed that mRNA is resistant to detergent/phenol extraction. This is thought to be due to the increased affinity between RNA and denatured protein caused by long tracts of poly(adenylic) acid.[13] This is particularly important when extracting tissues known to be very rich in protein, such as reticulocytes. The extraction and deproteinisation conditions can be altered to minimise this problem.[14] Sodium chloride is omitted from the detergent lysing medium and a low concentration (10 mM) of Tris buffer is added to produce alkaline conditions (pH 8.5). Chloroform, included in the deproteinising solution (chloroform : phenol mixture 1 : 1 by volume), decreases further the interaction between poly(A)-containing RNA and the protein-phenol interface, providing deproteinisation is carried out at room temperature. Nucleic acids are recovered from the aqueous layer by addition of NaCl (to 0.1M) and ethanol precipitation. A further method, by which the resistance of poly(A)-containing RNA to detergent/phenol extraction can be overcome, is to include a heat treatment to the protein-phenol interface. For example, re-extraction of the protein and phenol phases was carried out at 55°C to isolate globin mRNA from bone marrow cells.[15]

The affinity of poly(A)-containing RNA for denatured protein during phenol extraction can be more or less serious depending on the system concerned, but such considerations have led to the use, in certain cases, of proteinase K for mRNA extraction.[16,17] The basic principles behind this method were discussed earlier and are not significantly different for the isolation of mRNA.

Cell Fractionation

An alternative approach to the isolation of mRNA is to begin the preparation with cell fractionation. Although mRNA is now characterised in protein synthesising systems *in vitro* (see Chapter 7) mRNA can be defined as that RNA fraction that sediments with polyribosomes and can be released from polyribosomes by treatment with EDTA or puromycin. Polyribosome preparation has therefore been a commonly used first step in the isolation of mRNA. There are a number of advantages in beginning a mRNA preparation with the isolation of polyribosomes. First, the nuclear material (and therefore the bulk of DNA) is removed at an early stage and secondly, the majority of cellular protein is removed, thereby facilitating deproteinisation. There are, however, two considerable disadvantages in using polyribosomes as a source of mRNA. First, the initial isolation of polyribosomes can expose the mRNA to ribonuclease attack and secondly, only those mRNAs that are actually being translated at the time of tissue homogenisation can be isolated. The relative weight of these advantages and disadvantages varies from system to system. For example, polyribosome preparation works very well for the isolation of globin mRNA[18] but can allow ribonuclease degradation of ovalbumin mRNA.[19]

Polyribosomes are prepared initially by gentle lysis of the cells in near neutral buffer (for example 50 mM Tris-HCl pH 7.6 containing 0.15M sucrose) containing $MgCl_2$ (5–10 mM) and KCl (10–50 mM) to maintain ribosome stability.[20,21,22] A low concentration (0.1–0.3%) of a detergent such as SDS or Brij is often added to assist in the disruption of cytoplasmic membranes and diethylpyrocarbonate can be included to inhibit ribonuclease activity. Heparin has also been used to inhibit ribonuclease but not always with 100% success. It is likely that commercially available ribonuclease inhibitors may be increasingly used in polyribosome preparation. An essential detail in any polyribosome isolation is that the cell homogenate is kept ice-cold at all times.

The preparation of polyribosomes can be difficult if the cells possess a rigid cell wall. In the case of yeasts, spheroplasts are prepared prior to cell lysis[23] or in the case of plants, as gentle a homogenisation procedure as possible is used and extreme caution pertaining to ribonuclease inhibition is maintained. The yield of plant polyribosomes is increased if buffers of higher ionic strength than for the isolation of animal polyribosomes are used.[24]

Following cell lysis, nuclei, mitochondria and lysosomes are removed by centrifugation (12,000 g for 15 minutes) and polyribosomes are purified by centrifugation through sucrose. Continuous sucrose gradients are used to separate different size classes but discontinuous sucrose gradients or one-step sucrose cushion centrifugation are used for bulk polyribosome preparation. Polyribosome suspensions are opalescent, often with a faint yellow colour and are detected by absorbance at 260 nm. mRNA can be isolated from the polyribosome preparation by using a simple detergent solution (0.5% SDS for eukaryotic cytoplasmic ribosomes or up to 2% SDS for ribosomes from bacteria, chloroplasts and mitochondria) often with the inclusion of EDTA (10 mM) at pH 9.0 to disrupt the polyribosomes. Deproteinisation can then be carried out with an equal volume of phenol mixture and chloroform as described earlier.

If the mRNA to be isolated is well defined and the sedimentation constant (S) is known, it is possible to isolate the mRNA without using phenol or proteinase K. This has proved to be successful for the isolation of globin mRNA from rabbit reticulocyte lysate.[25] In this method, polyribosomes are disrupted in 0.5% SDS at 37°C and the RNA molecules separated in a 15% to 30% sucrose gradient. The fractions corresponding to mRNA (e.g. 9S for rabbit reticulocyte globin mRNA) are pooled and the nucleic acids directly precipitated with alcohol.

Affinity Chromatography

Eukaryotic messenger RNA can be separated from other RNA species by affinity chromatography. This is because the vast majority of mRNA molecules contain poly(adenylic) acid (20–250 bases) at the 3′ end, although histone mRNA is a known exception. Poly(A)-containing RNA was found to absorb onto cellulose nitrate filters and thus formed the basis of a partial purification procedure.[26] This method is not very specific, however, as non-specific absorption of rRNA can occur and at least 50 nucleotides of poly(A) per molecule is required for binding. A more efficient method has been developed that involves affinity chromatography on oligo (dT)-cellulose or poly(U)-Sepharose.[27] These materials are available commercially, consist of polymers of about 10–20 nucleotides and bind RNA containing a poly(A) tract as short as 20 bases.

Only very small columns are required since lg of oligo(dT)-cellulose will bind 20–40 A_{260} units of poly(A). A column containing 20 mg of oligo(dT)-cellulose is therefore adequate for most preparations. Poly(A)-containing RNA binds to the column at neutral pH in high salt concentrations (0.3M to 0.5M NaCl or KCl). Unbound nucleic acid can be removed by thorough washing in the same buffer. The mRNA is then recovered by simply eluting the column with neutral buffer containing no salt. The mRNA can then be precipitated from the elution buffer by the addition of NaCl (to 0.1M) and ethanol as described previously.

Inclusion of SDS in the chromatography buffers is a useful precaution against ribonuclease activity but this must be removed before *in vitro* translation of the mRNA. There is little to choose between oligo(dT) and poly(U) affinity ligands, but since the latter often contains slightly longer nucleotide tracts, stronger elution conditions, such as formamide or elevated temperature, may be necessary.

With either chromatography method a small proportion of the translatable mRNA activity does not bind to the column. This could be due to the poly(A) segments being too short or blocked in some way. A small percentage of rRNA or tRNA usually binds to the column but the effect of this can be reduced by repeating the affinity chromatography step.

Specific Methods

So far, the discussion of mRNA extraction has centred round the techniques for the isolation of total mRNA or mRNA species that represent a large proportion of the cellular RNA. It is essential, however, for techniques to be developed that facilitate the extraction of specific mRNA molecules which are invariably present in small numbers. This is a relatively new area and the principles involved will be briefly discussed here.

Polyribosome immunoprecipitation is a technique that can be developed for a wide range of systems and has been used to isolate specific mRNA that constitutes only 1% of the total mRNA population. The procedure consists of raising an antibody to the purified native protein coded by the mRNA of interest and is based on the premise that the antibody will recognise nascent protein chains.[28] Incubation of this antibody with a polyribosome preparaton does not, however, generally result in polyribosome precipitation. A second antibody raised against the first antibody is added to the incubation. The final complex forms a precipitate that is sufficiently dense to be separated from unreacted polyribosomes and antibody in a sucrose gradient. Alternatively, protein A (a commercially available, cell wall constituent of *Staphylococcus aureus* that binds to immunoglobulin G) can be used in place of the second antibody.[29] The mRNA is isolated from the antibody-polyribosome complex by detergent/phenol methods and oligo(dT)-cellulose chromatography. Polyribosome precipitations have been successfully used for a number of mRNAs including those for immunoglobulin from mouse myeloma, β casein from ewe mammary glands, vitellogenin from *Xenopus* liver and collagen from chick embryos.[29]

The availability of production of pure antibodies is an essential requirement of this procedure. Affinity chromatography with immobilised antigens has been used for antibody purification but it is likely that recent techniques of monoclonal antibody production will prove extremely useful in this respect. For a review of monoclonal antibodies the reader is referred to the article by Eisenbarth.[30]

A specific mRNA extraction procedure that is likely to become increasingly important involves the exploitation of DNA-RNA hybridisation methodology.[31] For example, virus mRNA from infected cells can be isolated by hybridisation of total cytoplasmic RNA to restriction fragments of virus DNA. Hybrids separated from the bulk RNA by hydroxyapatite chromatography are then denatured by heat and the mRNA separated from DNA by sucrose gradient centrifugation.

DNA sequences complementary to mRNA (cDNA) can be used in the purification and identification of mRNA. A pure mRNA preparation can be used as a template for cDNA synthesised directly onto oligo(dT)-cellulose. The poly(A) tail of the mRNA hybridises to the oligo(dT) which then forms a primer for the DNA polymerase stage of cDNA synthesis. The mRNA template can be removed by alkali treatment leaving an affinity column specific for a particular message. This technique has been successfully used for the isolation of the precursor to globin mRNA and ovalbumin mRNA.

Alternatively, synthetic oligodeoxyribonucleotides, based on knowledge of the mRNA base sequence or protein amino acid sequence, can be used as primers for cDNA synthesis, which can then be used in affinity columns as above or as hybridisation probes for specific mRNA fractionated on polyacrylamide or agarose gels ('Northern blotting'). For a more detailed discussion of the use of sequence-specific mRNA isolation procedures refer to the review by Taylor,[32] and the article by Goldberg *et al.*[33]

4. The Estimation of mRNA Content by Poly(U) Hybridisation

Although mRNA is generally characterised by translation in protein synthesising systems *in vitro* (see Chapter 7) it is often convenient to quantify the mRNA by estimating the poly(A) content. This can be done using a very sensitive technique for detecting poly(A) RNA that relies on hybridisation with high specific activity ^{3}H-poly(U). RNA is incubated with the ^{3}H-Poly(U) in 2 × SSC (SSC is 0.15M sodium chloride and 0.015M sodium citrate) at pH 7.0 for three minutes at 90°C, followed by an hour at 25°C to allow formation of hybrids.[34,35] Excess ^{3}H-poly(U) is then removed by treatment with ribonuclease A. Hybrids are collected by TCA precipitation and filtration and the radioactivity estimated in a scintillation counter. If the results are compared with data obtained from a similar experiment using standard poly(A), a value for the poly(A) content of unknown RNA preparations can be obtained. Poly(A) contents as low as 1–2 picograms can be estimated by this technique.

5. The Fractionation of RNA

Introduction

There are a number of different methods available for the fractionation of RNA. Some of the methods, for example affinity chromatography, are an integral part of many RNA extraction procedures and have already been described. Additional fractionation procedures are often required, however, to purify particular RNA molecules or to analyse the RNA product. This section will not attempt to review all the available fractionation procedures in detail but will instead concentrate on the few most generally applicable, presently used methods.

Solubility Methods

One of the simplest techniques for the fractionation of RNA is based on exploiting the differential solubility of nucleic acids in salt solutions. For example, high molecular weight RNA precipitates in strong solutions of sodium chloride (4M) and can therefore be relatively easily separated from low molecular weight RNA and DNA (and, incidentally, glycogen). Organic solvents can also be used to fractionate RNA. For example, rRNA can be precipitated from a total RNA solution (dissolved in 20% sodium benzoate and 3% sodium chloride) by the addition of 10% m-cresol. For further details of the application and use of solubility methods refer to the book by Parish.[36]

Ion-exchange Chromatography

Ribonucleic acids are highly charged molcules and bind to anion-exchange resins such as DEAE-cellulose. Binding is very strong and is directly proportional to the molecular weight of the RNA. For this reason extreme conditions of salt concentration or pH are required for the elution of high molecular weight RNA. The technique has therefore found most use in the fractionation of tRNA. By eluting an ion-exchange column with a salt solution it is possible to separate different tRNA species; a commonly used

Figure 6.1: Apparatus, made from perspex and platinum wire, for the electrophoresis of RNA in rod gels. The upper and lower tanks are filled with a suitable buffer and a voltage applied with a laboratory power pack.

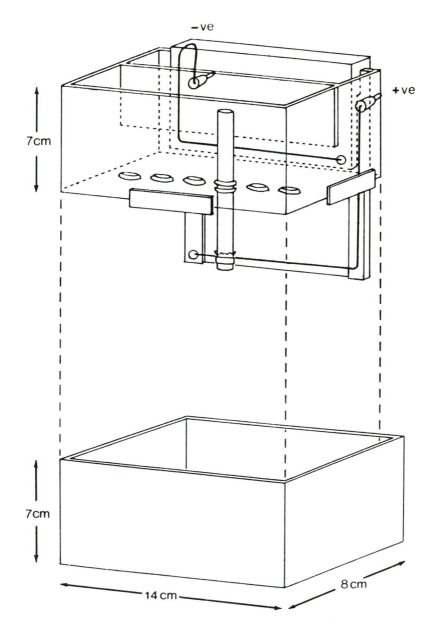

system is the elution of DEAE-cellulose with a linear concentration gradient of sodium chloride at acid pH in 4M urea. Further details of tRNA fractionation can be found in a series of articles edited by Grossman and Moldave.[37]

Sucrose Gradient Centrifugation

Centrifugation of RNA through a sucrose gradient is a fractionation technique that has been commonly used for the last twenty years. An RNA solution is layered over a previously prepared sucrose solution that increases in concentration towards the bottom of the centrifuge tube. Centrifugation of the tube results in separation of RNA into zones on the basis of sedimentation coefficient (S). Up to 5 mg of RNA can be fractionated per 50 ml tube in a large rotor and the technique can be adapted to zonal rotors.[38]

Sucrose gradient centrifugation is a convenient preparative fractionation procedure. The technique is useful for separating RNA species of high molecular weight (greater that 0.5×10^6 Daltons) and for separating high from low molecular weight RNA.[39] For example, rRNA from eukaryotes can be fractionated using a linear sucrose gradient (5–30% w/v) centrifuged overnight at approximately 80,000 g. Fractions can be collected from the gradient, RNA detected by monitoring the absorbance at 260 nm and the samples concentrated by ethanol precipitation. The disadvantages of sucrose gradients are that the resolution is relatively poor and ribonuclease activity during the fractionation can be a problem. The latter disadvantage can be overcome by treating solutions with diethylpyrocarbonate (see page 116). The poor resolution in sucrose gradients has led to the use of gel electrophoresis for the analytical fractionation of RNA and more recently for preparative fractionation.

Electrophoresis

RNA is a polyanion and thus migrates to the positive electrode during electrophoresis. If the migration is made to occur through a gel matrix of carefully chosen pore size, the mobility of the RNA molecules is approximately proportional to the logarithm of their molecular weight. A commonly used gel matrix for the fractionation of RNA is made up of polyacrylamide held in a suitable support such as glass or perspex tubes.[40] A diagram of a simple rod gel apparatus made from perspex is shown in Figure 6.1. The gels are formed by combining degassed acrylamide ($CH_2 = CHCONH_s$) and methylenebisacrylamide ($NH_2COCH = CHCH_2CH = CHCONH_2$) in solution with an initiator, 'TEMED' (tetramethylethylenediamine) and a catalyst, freshly prepared ammonium persulphate solution. The solution is poured into gel tubes (approximately 7 mm diamater) sealed at one end with dialysis membrane. Distilled water is then carefully layered over the acrylamide mixture to ensure a level upper surface and the gels are left to set for a couple of hours or overnight. The setting time and characteristics of gels are dependent to a certain extent on acrylamide concentration and ambient temperature.

After the gels have set, they are 'pre-run' by electrophoresis in neutral Tris-phosphate 'running' buffer containing EDTA and SDS for approximately 30 minutes. The RNA sample (up to 100 μg per gel) is then layered onto the top of the gel in running buffer (up to 100 μl) containing 10% sucrose. The power is then reapplied with a current of about 6 mA per tube (approximately 50 volts). During electrophoresis, the RNA will migrate in bands corresponding to the size of RNA, the smaller molecules moving the fastest.

Electrophoresis time is dependent on the type of RNA being fractionated and the resolution required. Electrophoresis times of between one and a half to four hours are common. Following electrophoresis the gels are carefully slid from the tubes. This is relatively easy if perspex tubes are used, but in the case of glass tubes, water has to be injected with a syringe between the gel and the tube before the gel can be removed.

RNA is best detected in the gels by monitoring the absorbance at 260 nm. This can be done with commercially available gel scanners. If no such device is available, the gel can be stained by soaking in 0.1% methylene blue (or toluidine blue) dissolved in 40% (v/v) aqueous 2-ethoxyethanol. Following staining (two hours) the gel is destained in 40% ethoxyethanol overnight.

Radioactivity, if present, can be detected by freezing the gel in solid CO_2, cutting the gels into slices (1 mm) with a gel slicer and counting radioactive emissions in a scintillation counter. If a weak β emitter such as ^3H is the radio-label, RNA must be eluted from the gel slices, before addition of scintillation fluid, by treatment with ammonia, hydrogen peroxide or a solubilisng agent such as 'Soluene', to ensure efficient counting.

The pore size of the gel can be adjusted to obtain optimum fractionation of the RNA under study. For example, gels containing 2.4% acrylamide (bis:acrylamide, 1:20) can be used for the fractionation of eukaryotic rRNA. Prokaryotic-type rRNA is best fractionated in a 2.6% gel (bis:acrylamide, 1:20) and small molcular weight RNA (4 and 5S RNA) in a 5% to 7% gel (bis:acrylamide, 1:40). Total RNA can be fractionated in 2.4% or 2.6% acrylamide by running the gels for a shorter time (Figure 6.2). A preparative apparatus based on electrophoresis in rod gels is commercially available (see Chapter 14 for details).

Rod gels for the fractionation of RNA are, in many cases, being replaced by the use of slab gels.[41] The important advantages of using slab gels are that:

(i) Many samples can be fractionated under identical conditions in a single gel.

(ii) Resolution can be superior to that obtained in rod gels.

(iii) A photographic copy of the gel can be easily obtained.

(iv) RNA molecules can be easily transferred to a support suitable for hybridisation experiments.

Figure 6.2: Fractionation of total nucleic acids (50 µg) from Zea mays roots by electrophoresis in a 2.6% polyacrylamide gel. Electrophoresis was for 2 hours at 6mA and 50 volts.

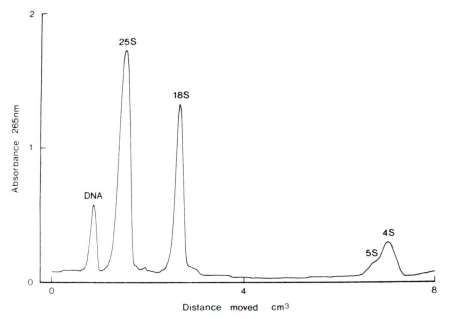

The disadvantage of slab gels is that it is less convenient to scan the gels with ultra-violet light. This means that staining procedures are generally used, thereby increasing the complexity of quantifying the RNA content of the separated bands.

High molecular weight RNA can be fractionated in agarose (1–2%) rather than acrylamide slab gels by using the apparatus and general procedure described for DNA fractionation (see Chapter 14) with the optional inclusion of SDS (0.5%) as a precaution against ribonuclease. For fractionation of lower molecular weight RNA acrylamide slab gels, stiffened with 0.5% agarose for ease of handling, are used. The fractionated nucleic acids can be visualised by methylene blue staining as previously described. Alternatively, RNA can be stained with ethidium bromide (provided that SDS is not present) and visualised by fluorescence in ultra-violet light. (Figure 6.3). Recently, it has been reported that the silver stain described for protein gels (see Chapter 2) also stains RNA in polyacrylamide gels with a sensitivity of 10 to 30 times that of ethidium bromide.[42] Radioactivity is most conveniently detected in slab gels by autoradiography.[43] Details of this technique are described elsewhere in this volume (see Chapter 15).

SDS is not particularly effective in denaturing RNA or disrupting aggregates. If accurate molecular weight values are required for the RNA, electrophoresis must be carried out in the presence of denaturing agents

Figure 6.3: Fractionation of total RNA (A: 2 μg, B: 5 μg) from Spinacea oleraceae *leaves by electrophoresis for 2½ hours at 40mA and 50 volts in a 1.4% agarose gel in the presence of 50% formamide. Following electrophoresis the gel was stained with ethidium bromide and viewed with ultra-violet light. Figures to the right of the gel represent the molecular weight (× 10⁻⁶) of the fragments.*

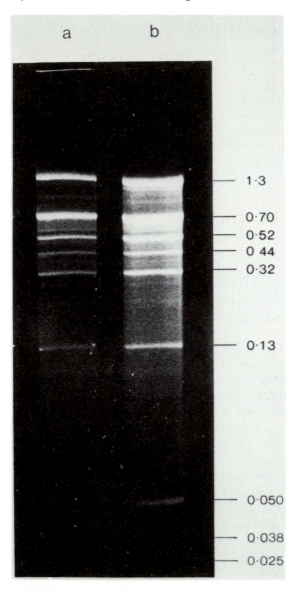

Source: Courtesy of S.N. Covey, John Innes Institute, Norwich, UK.

such as formamide, methyl mercury, formaldehyde or urea.[44,45] Aggregates of RNA can be disrupted by heating samples at 60°C in 8M urea prior to electrophoresis.

Recently, details of a technique using glyoxal treatment of nucleic acids followed by gel electrophoresis has been described for the accurate determination of nucleic acid molecular weights.[46] Glyoxal denatures RNA and DNA at neutral or acidic pHs by reacting with the nucleotide bases (in particular guanosine) and thus preventing the formation of base-pairs. Samples to be fractionated are heat-denatured in dimethyl sulphoxide and de-ionised glyoxal and can then be immediately loaded onto the gel. A convenient aspect of this technique is that denatured and untreated nucleic acids can be analysed on the same gel.

Acridine orange is the preferred stain for detection of nucleic acids fractionated by this method because ethidium bromide stains glyoxylated nucleic acids very poorly. Following destaining (one hour to overnight depending on conditions) the bands are visualised by illumination in ultra-violet light ($\lambda = 254$ nm). With this staining technique double-stranded nucleic acids appear green and single-stranded nucleic acids generally appear red or orange. If the gel is sufficiently destained, 0.1 µg of single-stranded and 0.05 µg of double-stranded nucleic acids can be detected in each band. If necessary resolution in slab gels can be improved by using a two-dimensional electrophoresis procedure as recently described for the separation of 4-12S RNA.[47]

RNA fractionated on slab gels can be characterised by hybridisation to specific probes such as ^{32}P-labelled cDNA made by nick-translation (see Chapter 15). In this technique RNA is transferred from the gel to nitrocellulose filters under high salt conditions[48] or to diazo-benzyloxymethyl (DBM) paper.[49] This technique (referred to as 'Northern blotting') is an adaptation of the DNA transfer technique ('Southern blotting') described in Chapter 15.

Although gel electrophoresis is most commonly used as an analytical method, the technique can be used as a preparative procedure. RNA can be eluted (in high salt buffer containing SDS) or electrophoresed out of polyacrylamide gels but the product is usually contaminated with acrylamide.[50] For this reason, agarose gels are often preferred for preparative electrophoresis[51] and the technique is therefore essentially the same as that described for DNA. Details of preparative gel electrophoresis are given in Chapter 14.

References

1. J.H. Parish, *Principles and practice of experiments with nucleic acids*, (Longman, 1972).
2. K.S. Kirby, 'Isolation of nucleic acids with phenolic solvents', *Methods in Enzymology*, *12B* (1968), 87–99.
3. Parish, *Principles and practice*.
4. C.J. Leaver and J. Ingle, 'The molecular integrity of chloroplast ribosomal nucleic acid', *Biochem. J.*, *123* (1971), 235–244.
5. Parish, *Principles and practice*.
6. D. Gillespie, 'The formation and detection of DNA-RNA hybrids', *Methods in Enzymology*, *12* (1968), 641–668.
7. U. Weigers and H. Hilz, 'Rapid isolation of undegraded polysomal RNA without phenol', *FEBS Letts*, *23* (1972), 77–82.
8. H. Hilz, U. Wiegers and P. Adamietz, 'Stimulation of proteinase K action by denaturing agents: appliction to the isolation of nucleic acids and the degradation of "masked proteins"', *Eur. J. Biochem.*, *56* (1975), 103–108.
9. Weigers and Hilz, *Rapid isolation of undegraded polysomal RNA*.
10. R.A. Cox, 'The use of guanidinium chloride in the isolation of nucleic acids', *Methods in Enzymology*, *12B* (1968), 120–129.
11. V. Glisin, R. Crkvenjakov and C. Byus, 'Ribonucleic acid isolated by caesium chloride centrifugation', *Biochemistry*, *13* (1974), 2633–2637.
12. A.W. Nienhuis, A.K. Falvey and W.F. Anderson 'Preparation of globin messenger RNA', *Methods in Enzymology*, *30* (1974), 621–630.
13. G. Brawerman, 'The isolation of messenger RNA from mammalian cells', *Methods in Enzymology*, *30* (1974), 605–612.
14. Ibid.
15. Nienhuis, Falvey and Anderson, *Preparation of globin messenger RNA*.
16. Weigers and Hilz, *Rapid isolation of undegraded polysomal RNA*.
17. D.D. Brown and Y. Suzuki, 'The purification of the messenger RNA for silk fibroin, *Methods in Enzymology*, *30* (1974), 648–654.
18. Nienhuis, Falvey and Anderson, *Preparation of globin messenger RNA*.
19. J.M. Rosen, S.L.C. Woo, J.W. Holder, A.R. Means and B.W. O'Malley, 'Preparation and preliminary characterisation of purified ovalbumin messenger RNA from hen oviduct', *Biochemistry*, *14* (1975), 69–78.
20. Parish, *Principles and practice*.
21. C.E. Sripati and J.E. Warner, 'Isolation, characterisation and translation of mRNA from yeast, *Meth. in Cell Biol.*, *20* (1978), 61–81.
22. R.T. Schimke, R. Palacios, D. Sullian, M.L. Kiely, C. Gonzales and J.M. Taylor, 'Immunoadsorption of ovalbumin synthesising polysomes and partial purification of ovalbumin messenger RNA', *Methods in Enzymology*, *30* (1974), 631–648.
23. Sripati and Warner, *Isolation, characterisation and translation*.
24. E. Davies, B.A. Larkins and R.H. Knight, 'Polyribosomes from peas: an improved method for their isolation in the absence of ribonuclease inhibitors', *Plant Physiol.*, *50* (1972), 581–584.
25. Nienhuis, Falvey and Anderson, *Preparation of globin messenger RNA*.
26. Brawerman, *The isolation of messenger RNA*.
27. H. Nakazato and M. Edmonds, 'Purification of messenger RNA and heterogeneous nuclear RNA-containing poly(A) sequences', *Methods in Enzymology*, *29* (1974), 431–443.
28. Schimke, Palacios, Sullivan, Kiely, Gonzales and Taylor, *Immunoadsorption*.
29. J.M. Taylor 'The isolation of eukaryotic messenger RNA', *Ann. Rev. Biochem.*, *48* (1979), 681–717.
30. G.S. Eisenbarth, 'Application of monoclonal antibody techniques to biochemical research', *Anal. Biochem.*, *111* (1980), 1–16.
31. Taylor, *The isolation of eukaryotic messenger RNA*.
32. Ibid.
33. M.L. Goldberg, R.P. Lifton, G.R. Stark and J.G. Williams, 'Isolation of specific RNAs using DNA covalently linked to diazobenzyloxymethyl cellulose or paper', *Methods in Enzymology*, *68* (1979), 206–220.

34. S.N. Covey and D. Grierson, 'The measurement of plant polyadenylic acid by hybridisation with radioactive polyuridylic acid', *Planta, 131* (1976), 75–79.
35. J.O. Bishop, M. Rosbash and D. Evans, 'Polynucleotide sequences in eukaryotic DNA and RNA that form ribonuclease-resistant complexes with polyuridylic acid', *J. Molec. Biol., 85* (1974), 75–86.
36. Parish, *Principles and practice*.
37. L. Grossman and K. Moldave (eds.), 'Nucleic acids and protein synthesis. Part E', *Methods in Enzymology, 29* (1974).
38. Parish, *Principles and practice*.
39. E.H. McConkey, 'The fractionation of RNAs by sucrose gradient centrifugation', *Methods in Enzymology, 12A* (1967), 620–634.
40. Parish, *Principles and practice*.
41. R. De Wachter and W. Fiers, 'Fractionation of RNA by electrophoresis on polyacrylamide slab gels, *Methods in Enzymology, 21* (1971).
42. M.J. Berry and E.C. Samuel, 'Detection of subnanogram amounts of RNA in polyacrylamide gels in the presence and absence of protein by staining with silver', *Anal. Biochem., 124* (1982) 180–184.
43. M.W. Schwinghamer and R.J. Shepherd, 'Formaldehyde-containing slab gels for analysis of denatured, tritium-labelled RNA', *Anal. Biochem., 103* (1980), 426–434.
44. H. Lehrach, D. Diamond, J.M. Wozney and H. Boedtker, 'RNA molecular weight determinations by gel electrophoresis under denaturing conditions, a critical re-examination', *Biochemistry, 16* (1977), 4743–4751.
45. T. Maniatis and A. Efstratiadis, 'Fractionation of low molecular weight DNA or RNA in polyacrylamide gels containing 98% formamide or 7M urea, *Methods in Enzymology, 65* (1980), 299–305.
46. G.G. Carmichael and G.K. McMaster, 'The analysis of nucleic acids in gels using glyoxal and acridine orange, *Methods in Enzymology, 65* (1980), 380–391.
47. K. Takeishi and S. Kaneda, 'A two dimensional gel electrophoretic procedure for the separation of complex mixtures of 4-12S RNAs', *Anal. Biochem., 113* (1981), 212–218.
48. P.S. Thomas, 'Hybridisation of denatured RNA and small DNA fragments transferred to nitrocellulose', *Proc. Nat. Acad. Sci. USA, 77* (1980), 5201–5205.
49. J.C. Alwine, D.J. Kemp, B.A. Parker, J. Reiver, J. Renart, G.R. Stark and G.M. Wahl, 'Detection of specific RNAs or specific fragments of DNA by fractionation in gels and transfer to diazobenzyloxymethyl paper, *Methods in Enzymology, 68* (1979), 220–244.
50. Parish, *Principles and Practice*.
51. J. Langridge, P. Langridge and P.L. Bergquist, 'Extraction of nucleic acids from agarose gels', *Anal. Biochem., 103* (1980), 264–271.

7 PROTEIN SYNTHESIS *IN VITRO*

Don Grierson and Jim Speirs

CONTENTS

1. Introduction

There are two main reasons for carrying out *in vitro* protein synthesis, or translation. Firstly, to provide information about the mechanism of protein synthesis and the identity and structure of the reacting components. Secondly, to identify mRNA molecules and to study their properties and the properties of the proteins for which they code. The first objective involves the purification and characterisation of each component and a study of its role in protein synthesis. This represents an entire field of study in itself and will not be discussed in detail. Further information may be obtained from recent accounts of protein synthesis.[1,2] The second objective can be achieved by making a crude extract, from a suitable biological source, which contains a complete mixture of all the components necessary for protein synthesis. This extract is then used for *in vitro* translation of purified fractions containing mRNA (Figure 7.1).

Functional protein-synthesising systems have been prepared from extracts of a variety of animals, plants and microorganisms. Most of the information about the biochemistry of protein synthesis has come from studying and analysing the composition and activity of individual components of such extracts. Although a few small peptides are synthesised without a template, using enzymes alone, the synthesis of most proteins requires the following subcellular components: ribosomes (either of the prokaryotic '70S' type or the eukaryotic '80S' type); a complete mixture of tRNAs; amino-acyl-tRNA-synthetases (activating enzymes); enzymes for polypeptide chain initiation, elongation and termination; energy, in the form of ATP and GTP; cations such as K^+ and Mg^{2+} and a template or mRNA.

When mRNA is added to an *in vitro* protein synthesis system containing the necessary components, the protein coded by the mRNA is synthesised. Thus, *in vitro* protein synthesis makes it possible to identify and quantify mRNA molecules by analysing their protein products. Using this approach, it is possible to study changes in mRNA content during growth and development, or in response to hormones. *In vitro* translation systems may also be used, in a variety of ways, to identify cloned genes or mRNA sequences required for cloning, and also to study post-translational modification of proteins.

Figure 7.1:   General Scheme for in vitro *Protein Synthesis Experiments*

Preparation of <u>in vitro</u> translation system from <u>E. coli</u> wheat germ or rabbit reticulocytes (see Section 3)

Extract and purify RNA (see Chapter 5) Total cell RNA or polysomal RNA or Poly(A)-containing RNA or RNA from chloroplasts or mitochondria

Autoclave solutions and buffers before use and use aseptic techniques where possible. Store the extract as small frozen aliquots, preferably in liquid nitrogen. The extract should contain very little endogenous mRNA and low concentrations of the amino acid to be used as radioactive tracer

Prevent ribonuclease digestion of RNA during extraction and purification. Remove DNA, protein, carbohydrate, etc and store RNA as a precipitate under ethanol (see Chapter 5)

Thaw out sufficient volume of extract but keep in ice bath. Add radioactive amino acids and any other required components from prepared stock solutions (sterile) (see Section 3)

Each incubation is carried out in a final volume of 10–50μl. Add appropriate volume of RNA solution Incubate at the required temperature for 30–90 mins (see Section 4)

Remove ions and detergents by washing RNA pellet several times with 80% ethanol Remove ethanol by evaporation, dissolve RNA in sterile distilled water and store frozen

Stop the reaction, measure the incorporation of radioactive amino acids into protein and fractionate the radioactive proteins for further study (see Section 4)

Notes: Whenever possible use sterile glassware and solutions, and take care not to introduce <u>any</u> contaminant. Do not repeatedly freeze and thaw the <u>in vitro</u> protein synthesis extract. Protein synthesis is very sensitive to pH and inorganic ion concentrations. Therefore, pipette accurately using micropipettes. The reaction is adversely affected by traces of detergent, phenol, ions such as Ca^{2+} and Na^+, ethanol and contaminants in the RNA preparation which alter the pH. Purify the RNA thoroughly and use sterile distilled water of the highest purity available.

2. Basic Requirements for *in vitro* Translation of mRNA

Choosing the in vitro *Translation System*

Ideally an *in vitro* translation system should be inexpensive, easy to prepare, stable during storage, capable of translating mRNA molecules from a variety of sources and have a low background of translation of endogenous mRNA. The *in vitro* translation systems available are derived from cell extracts which, during preparation, are either partially fractionated or more completely fractionated and reconstituted (see Section 3). They contain all the factors necessary for protein synthesis mentioned above, and fulfil the main criteria for a translation system which are a low level of endogenous mRNA, very low concentrations of the amino acid to be used as radioactive tracer (see below) and efficient and correct translation of added mRNA.

In general, mRNA purified from any eukaryotic organism (i.e. one containing an '80S' ribosome system) can be translated in an *in vitro* protein synthesising system derived from any other eukaryote. Thus, animal mRNAs can be translated in wheat germ extracts and plant mRNAs can be translated in rabbit reticulocyte lysates (Table 7.1). This compatibility exists because these organisms all share the same genetic code (with the exception of at least some mitochondria) and have similar nucleotide sequences for recognition and binding of mRNA to ribosomes and other components of the protein synthesising apparatus. However, a few exceptions to this generalisation have been found. For example, large mRNAs may not be translated very efficiently in wheat germ extracts because of premature polypeptide chain termination. Furthermore, occasional instances have been noted where a particular mRNA is preferentially translated in a wheat germ extract but not in a reticulocyte lysate, and vice versa. This may be due to subtle differences in recognition sequences or translational control mechanisms in different eukaryotes.

It is widely believed that the nucleotide recognition sequences of mRNAs and ribosomes (but not the genetic code) from prokaryotic organisms and prokaryotic-like organelles of eukaryotes are so different from those of the 80S cytoplasmic system of eukaryotes that mRNA from one type of organism or organelle can not be translated in a protein synthesising system derived from the other. This is probably a good general rule to bear in mind when choosing a suitable *in vitro* translation system (Table 7.1) but there are interesting exceptions. For example, *Euglena* and wheat chloroplast mRNA can be translated in a wheat germ system[3,4] and spinach chloroplast mRNA has been translated in a rabbit reticulocyte lysate.[5,6]

In order to prepare *in vitro* translation systems in the laboratory, it is either necessary to culture cells, maintain laboratory animals or in the case of the wheat germ extract, to test several sources for the most suitable starting material (see Table 7.1 and Section 3). Although the methods have been streamlined as much as possible, they are still time-consuming and require access to a reasonably well equipped laboratory. All the preparations are relatively unstable and lose biological activity unless stored at low temperatures, either in a −80°C deep-freeze or, preferably, under liquid nitrogen.

When large numbers of *in vitro* translation experiments are being carried out, there is obviously a great financial saving to be made if the translation system is prepared in the laboratory. However, if the appropriate facilities are not available, or if difficulty is experienced in preparing an active extract, it may be worthwhile to purchase a commercial preparation of wheat germ extract or reticulocyte lysate. Although the cost appears high, there is an important saving in preparation time and in ancillary materials.

Table 7.1: Choosing the in vitro System

	Source of mRNA	Suitable System
a)	Prokaryotic organisms such as bacteria, bacteria infected with 'phage and prokaryote-like organelles such as chloroplasts and some mitochondria.	*E. coli* S-30 extract[a]
b)	Eukaryotic organisms such as yeast, algae, higher plants, protozoa, invertebrate and vertebrate animals including mammals, and viruses which infect these organisms.	Rabbit reticulocyte lysate[b,c] or Wheat germ extract[d] or *Xenopus* oocytes[e]

Notes: a. *E. coli* S-30 extract. This is relatively simple to prepare. A ribonuclease-deficient strain of *Escherichia coli* and a French pressure cell are an advantage (see Section 3).

b. Rabbit reticuloycte lysate. This is simple to prepare *providing* you have access to rabbits and the help of a licenced person to bleed the animals. The performance of the system is substantially improved if pre-treated with micrococcal nuclease to remove endogenous mRNA (see Section 3). Very high rates of protein synthesis can be obtained and large mRNAs are efficiently translated.

c. Commercial preparations. It is now possible to purchase rabbit reticulocyte lysate from several firms including Amersham International and New England Nuclear Enterprises. Wheat germ extract can be purchased from Bethesda Research Laboratories Inc. Both types of preparation are relatively expensive but they come almost ready to use and they can save a lot of time.

d. Wheat germ extract. This is relatively easy to prepare but it is essential to obtain a good (non-roasted) wheat germ, preferably from a hard winter wheat. The wheat germ should not be contaminated with endosperm, which is a potential source of nucleases and proteases. Several chemicals have to be added to the final extract and this can increase the preparation time. Endogenous mRNA activity is relatively low. Large mRNAs may not be translated very efficiently.

e. *Xenopus* oocyte. Frogs eggs are not, strictly speaking, an *in vitro* system, although they are very useful for some experiments. It is necessary to maintain a colony of *Xenopus* to provide the oocytes and it takes some skill and special equipment to micro-inject RNA into the oocytes. The advantages are that only extremely small quantities of RNA are required and that processing of proteins can occur after translation (see Sections 3 and 4).

Selecting the Radioactive Amino Acid

Only very small quantities of protein are produced by *in vitro* translation of mRNAs and it is very rare for the protein products to be detected by their biological activity using enzyme assays. Indeed, in some cases no native protein is produced at all but a precursor protein is formed which requires specific modification by a post-translational mechanism (see Section 4). It is therefore necessary to follow the incorporation of radioactive amino acids both as a measure of protein synthesis and in order to identify specific protein products. The amino acid should be supplied at as high a specific radioactivity as possible in order to maximise the incorporation into protein. Several points have to be considered: the amino acid composition of the products (if known); the amino acid pool sizes of the *in vitro* translation system and the method of detection to be employed.

Radioactive amino acids containing either ^{35}S, ^{14}C or ^{3}H can be used for *in vitro* translation studies. These emit β particles of different energies, which affects the ease with which they can be detected. In many cases radioactive translation products are fractionated by gel electrophoresis and detected with X-ray film by autoradiography or fluorography (see Section 4). These methods detect ^{35}S much more rapidly than either ^{14}C or ^{3}H. The ^{35}S-containing amino acids (methionine and cysteine) are, therefore, the obvious choice, particularly as they are now available commercially at very high specific radioactivities, of the order of 10^3Ci mMole^{-1}. This is far in excess of the specific radioactivity of ^{3}H- or ^{14}C-labelled amino acids. Thus ^{35}S-methionine generally gives much higher incorporation of radioactivity into protein than ^{14}C- or ^{3}H-amino acids, even though methionine occurs relatively infrequently in many proteins. Another advantage is that the endogenous pool size of methionine in the reticulocyte lysate is relatively low and does not greatly dilute the added ^{35}S-methionine. (This is not a consideration with wheat germ extracts, see Section 3.) In cases where ^{35}S-methionine is not suitable, because the protein being studied contains little or no methionine, ^{35}S-cysteine may be a better choice than ^{3}H- or ^{14}C-labelled amino acids, for the reasons outlined above. However, it is necessary to prevent spurious binding of ^{35}S-cysteine to proteins in the translation mixture (see Section 4).

Recently, Amersham International Ltd have introduced a new product, L-^{75}Se-selenomethionine. This shows good incorporation into protein both *in vivo* and *in vitro* in reticulocyte lysates. It has the useful property of being a γ-emitter (see Section 4).

Preparing the RNA

Since the object of carrying out *in vitro* protein synthesis is to study mRNA, it makes sense to free it from other types of RNA before use. However it is not absolutely essential to purify the mRNA completely and a variety of different preparations can be used successfully for *in vitro* translation (Figure 7.2). The main requirement is that the RNA should be free of DNA, protein and carbohydrate, which can adversely affect *in vitro* translation of mRNA. Methods for the extraction and purification of mRNA are described in Chapter 6.

RNA preparations can be safely stored as ethanol-precipitates in a deep-freeze (−20°C or lower) for several weeks. To prepare the RNA for *in vitro* translation it should be pelleted, by low-speed centrifugation, and the tube containing the pellet should be drained. The pellet should then be washed, to remove traces of phenol, detergents, inorganic ions and other impurities, by resuspending the RNA in 80% ethanol, shaking the suspension and repelleting the RNA by low speed centrifugation. This procedure should be repeated at least twice. Care should be taken during this, and all subsequent steps, not to introduce contaminants into the RNA preparation. The main danger is the introduction of stray nucleases. It is best to use sterile

Figure 7.2: Scheme for the Isolation of various mRNA-containing Fractions for use in in vitro Protein Synthesis

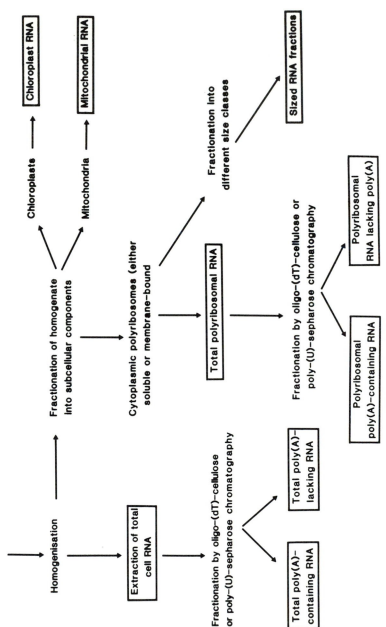

glassware and water, to keep tubes sealed where possible and to wear clean disposable gloves. The tubes containing the washed RNA pellets should be drained and dried for a few minutes in a vacuum desiccator, to remove ethanol. Drying can be speeded up if the RNA is given a final wash in 95% ethanol before pelleting. Finally, the RNA should be dissolved in a small volume of sterile distilled water at a concentration within the range 1–5 μg μl^{-1} and any insoluble material removed by centrifugation. Distilled water of high purity should be used to ensure the absence of inhibitory ions (see Section 4). The concentration of the RNA can be measured by withdrawing an aliquot with a microsyringe and measuring the u.v. spectrum. This gives an indication of the purity of the RNA and the concentration: an RNA solution with an optical density of 0.1 at 260 nm in a 1 cm light path has a concentration of approximately 4 μg ml^{-1}. The RNA solution can be used to direct *in vitro* protein synthesis immediately or it can be stored in a deep freeze (−20 or −80°C).

Carrying out in vitro *Translation Experiments*

Most of the effort is concerned with the preparation of the translation system and the RNA solution. The *in vitro* translation itself only involves a little pipetting. Each incubation is generally carried out in a final volume of between 10 and 50 μl. Pipetting is carried out with microsyringes and the reaction mixtures are incubated in very small stoppered glass or plastic tubes. When using a commercial reticulocyte system only three solutions are necessary: the lysate, the radioactive amino acid and the RNA (see Section 3). However, preparing a home-made wheat germ extract is a little more complicated and several other stock solutions are required (see Section 3).

3. Types of Translation System

Translation of mRNAs from Prokaryotes

The first cell-free synthesis of polypeptides was demonstrated by Hoaglend *et al.* in 1958 and, in 1961 Nirenberg and Mathaei stimulated the synthesis of polyphenylalanine by the addition of the synthetic polynucleotide poly(U) to an S-30 extract of *Escherichia coli*. ('S-30' stands for the supernatant fraction of lysed *E. coli* cells centrifuged at 30,000 g.)

Most systems used for translation of prokaryotic-type mRNAs are derived from *E. coli* S-30 extracts although extracts of a few other prokaryotes have been used, e.g. *Bacillus subtilis*[7] and *Halobacter*.[8]

Preparation of the E. coli *S-30 system.*   Reasonably consistent incorporations are obtainable using a modification of the protocol of Bottomley and Whitfield[9] which is a variation of that of Zubay *et al.*[10] A ribonuclease-deficient strain of *E. coli*, such as MRE 600 or PR 7, should be used for preparation of the lysate and care should be taken to ensure that the cells are harvested during vigorous growth, i.e. mid log-phase of a rapidly dividing culture. The cells should be rapidly chilled during harvesting and care should be taken during subsequent steps to keep the temperature below 4°C.

Cell lysis can be accomplished either by passage through a French pressure cell, or by grinding in alumina powder, although the former is easier. It is important to ensure the presence of a reducing agent, either mercaptoethanol or dithiothreitol, or a mixture of both, during and after lysis to prevent the oxidation of SH groups. The reducing agent must be added immediately prior to use as it is unstable.

The lysate is cleared by centrifugation at 30,000 *g*, and incubated under conditions designed to remove ribosomes from endogenous messenger RNAs and hence encourage their degradation by RNAases in the preparation. The length of incubation should be optimised.[11] The lysate is then desalted either by passage through a column of Sephadex G-25 or by dialysis. In our experience dialysis is preferable.

With some preparations it appears that excessive dialysis can be detrimental to the activity of the lysate and it may be necessary to optimise the dialysis conditions, i.e. length of dialysis and number of changes, in order to obtain the highest possible activity. In our laboratory, overnight dialysis with four changes of dialysate produced the most active preparations.

A convenient way of storing the active S-30 fraction is to drop it slowly from a Pasteur pipette into liquid nitrogen where it forms frozen balls with a volume of approximately 50 µl. The balls can be stored over liquid nitrogen where they will retain their activity for several months. Storage at −80°C also preserves the activity of the preparation but activity is rapidly lost at − 20°C or higher.

The final S-30 extract used for *in vitro* protein synthesis contains:

mRNA fraction	Added at a concentration optimised for each assay – about 100–600 µg ml^{-1}
ribosomes initiation elongation termination } factors amino acids transfer RNA including fMet tRNA	Present in the S-30 extract
labelled amino acid	Added at a concentration of from 20–100 µ Ci ml^{-1}
GTP ATP Phosphoenol pyruvate Pyruvate kinase Reducing agent	Added to the extract
Inorganic salts	Some present in the extract and others added

RNA-directed protein synthesis requires a slightly different reaction mixture to those described in references by Bottomley and Whitfeld[12] and Zubay *et al.*[13] as these deal with DNA-directed synthesis. Preparation of the assay mixture is described in detail by Modolell[14] and a modified system is described by Speirs and Grierson.[15]

In theory, desalting of the *E. coli* lysate should lower the concentration of amino acids and it is standard practice to add unlabelled amino acids to the translation mixture (*excluding* the amino acid to be used as label). In practice it appears that the amino acids are not fully depleted and, while this is not important when considering whether or not to add amino acids, it makes it difficult to calculate in absolute terms the activity of the added labelled amino acid. Exhaustive desalting or dialysis will not fully overcome this problem as there will always be a pool of unlabelled amino acids present as amino-acyl tRNAs, which will not be removed by desalting or dialysis.

ATP sequesters Mg^{2+} ions and thus at high concentrations inhibits protein synthesis. Therefore, the energy requirements are provided in part by ATP and in part by a system for converting ADP to ATP. Phosphoenol pyruvate is used as the source of phosphate and in many systems pyruvate kinase is added to catalyse the $ADP \rightarrow ATP$ reaction. Care must be taken, however, as many commercial preparations of pyruvate kinase have an RNase contaminant. It will be found that most *E. coli* S-30 preparations have sufficient endogenous pyruvate kinase activity and good incorporation levels can be obtained by adding only the phosphoenol pyruvate without the kinase.

The salt requirements of the *E. coli* system are not critical although they should be optimised for each new preparation. Reducing agents used should be freshly made up.

E. coli lysates do contain low levels of ribonuclease and endogenous *E. coli* mRNAs tend to be inactivated during preparation and pre-incubation of the S-30 fraction. Should the level of active endogenous mRNA prove a problem during the translation of exogenous mRNAs, this can be overcome in one of two ways. The S-30 can be made up in an assay mixture containing everything except exogenous mRNA and labelled amino acid and can be briefly pre-incubated (five minutes at 37°C). This normally results in an appreciable drop in endogenous mRNA activity with only a small loss in efficiency of exogenous mRNA translation. A more efficient technique is the inactivation of endogenous mRNAs by incubation with micrococcal nuclease in the presence of calcium ions, following the procedure outlined by Pelham and Jackson[16] for the rabbit reticulocyte system. The micrococcal nuclease, which is calcium-dependent, is inactivated prior to the addition of exogenous mRNA by chelating the calcium ions with EGTA. Care must be taken to add an excess of EGTA as calcium is very inhibitory to protein synthesis and free calcium activates the nuclease and leads to degradation of added mRNA.

Translation of mRNAs from Eukaryotes

Eukaryotic *in vitro* translation systems have been derived from a large number of diverse sources. The most commonly used systems are those obtained from wheat germ[17,18] and from rabbit reticulocytes.[19,20] However, systems have also been derived from Krebs ascites II cells,[21] wheat embryo,[22] yeast,[23] rye embryo,[24] mung bean primary axes,[25] pea primary axes,[26] Chinese hamster ovary cells[27] and other sources.

Most of these systems are very similar and for the purposes of discussing eukaryotic *in vitro* protein synthesis we will consider only the wheat germ and the rabbit reticulocyte lysate systems.

The Wheat Germ System. The preparation of the wheat germ lysate fraction (the S-30 fraction) containing the required components for *in vitro* translation is very well outlined by Roberts and Patterson[28] and Marcu and Dudock.[29] We have found the latter method consistently gives greater activities. A brief outline of the procedure is:

(i) Grind a few g of wheat germ in a cooled mortar and pestle, adding glass fragments or beads to help disrupt the tissue.

(ii) Add a few ml of chilled extraction buffer and rapidly homogenise to a thick paste.

(iii) Transfer the mixture to a centrifuge tube with a spatula and spin at 15–30,000 g for 15 min at 4°C.

(iv) Take the supernatant liquid, avoiding the pellet and the floating fatty layer. Pass the extract down a Sephadex G-25 column, pre-equilibrated with buffer, in a cold-room and collect the first 2 or 3 ml of the excluded (cloudy) fraction. This has a very high A_{260} nm.

(v) Centrifuge the column-fraction at 15–30,000 g for 15 min at 4°C.

(vi) Add the solution drop by drop to liquid nitrogen and freeze it as small balls. Store the extract over liquid nitrogen.

The wheat germ used for preparation of *in vitro* translation systems should preferably be obtained from a hard winter wheat. These are not generally available in Britain and Europe and it is necessary to obtain material either from laboratories using the wheat germ system or directly from the grain mills. If the wheat germ to be used can be seen to be contaminated with endosperm, this can be removed by flotation on cyclohexane/carbon tetrachloride.[30]

The prepared S-30 fraction can be stored in the same way as the *E. coli* S-30 fraction by freezing droplets in liquid nitrogen and storing them above liquid nitrogen or at −80°C. It is necessary to add the following components before use: unlabelled amino acids (excluding the one to be used in labelled form), the radioactive amino acid, polyamines (spermine and/or spermidine), components for energy production (ATP, GTP, creatine phosphate and, in some cases, creatine kinase) and optimum concentrations of K^+ and Mg^{2+}.

The use of the wheat germ system is described in detail by Marcu and Dudock.[31] Wheat germ is an ideal source of ribosomes because of the low level of endogenous mRNA activity. We have noted on a few occasions that the addition of chloroplast tRNA to the wheat germ assay has increased the background levels of incorporation (unpublished results). It is possible this results from a stimulation of translation of endogenous mRNA in the wheat germ preparation. Wheat germ extracts can be freed of endogenous mRNA by treatment with calcium-dependent nuclease, as described for the *E. coli* and rabbit reticulocyte extracts.

The Rabbit Reticulocyte Lysate.   The most efficient eukaryotic *in vitro* translation system is probably that derived from lysed rabbit reticulocytes. It is more efficient in synthesising high-molecular-weight proteins than the wheat germ extract and each mRNA molecule is translated more times. Preparation and use of the system have been described in detail by Schimke *et al.*[32] and Hunt and Jackson.[33]

As prepared, the reticulocyte lysate contains a large pool of endogenous mRNA. Exogenous mRNA will only be translated if added in sufficient amounts to compete with the endogenous pool. This problem can be circumvented by degrading the endogenous mRNA with calcium-dependent micrococcal nuclease. The nuclease is then inactivated by chelation of Ca^{2+} by EGTA before addition of exogenous mRNA. However, it is important to guard against the introduction of Ca^{2+} to the lysate as this will activate the micrococcal nuclease, leading to the degradation of mRNA.

Commercial preparations of reticulocyte lysate are now available (from Amersham International and New England Nuclear Enterprises). Amersham International also produce a fractionated lysate which has a lower amino acid pool and consequently gives higher incorporation of radioactivity with ^{35}S-methionine. It is also easy to use alternative labelled amino acids with the fractionated lysate.

Xenopus Oocytes.   The injection of mRNA into *Xenopus* oocytes or eggs, to stimulate synthesis of protein, is not strictly *in vitro* protein synthesis but it is frequently used to achieve the same aims and objectives as *in vitro* translation systems.

The techniques used for the microinjection of solutions into *Xenopus* oocytes and eggs are described in detail by Lane *et al.*,[34] Gurdon *et al.*[35] and Stephens *et al.*[36]

The *Xenopus* system has a number of inherent problems. The need for a supply of oocytes or eggs requires the maintenance of a colony of *Xenopus laevis* frogs in the laboratory. The microinjection technique requires a sophisticated piece of equipment and some skill. Only minute quantities of RNA can be translated in a single egg, so that limited quantities of product are available for analysis. Furthermore, analysis and interpretation of results is complicated by the fact that the oocyte is actively synthesising its own spectrum of proteins. This problem can be overcome by studying a recognisable, non-*Xenopus*, protein product but it effectively limits the system to the translation of one or a few mRNAs at one time.

Although the *Xenopus* system does present problems to the potential user it has one major advantage over the other systems described here. Many proteins normally undergo 'processing' and packaging *in vivo*. Only partial processing and packaging has been achieved in the wheat germ and reticulocyte lysate systems[37] and this may require the addition of specially-prepared endoplasmic reticulum fractions (see Section 4). The *Xenopus* system, on the other hand, appears to be able to process heterologous proteins which have been translated from exogenous mRNA. In addition, packaging[38,39] and secretion of certain proteins has been demonstrated.[40]

S-100 in vitro *Translation Systems*

The S-100 system (also known as S-150, S-137, etc.) is an *in vitro* protein synthesis extract, from either prokaryotic or eukaryotic cells, from which the ribosomes have been removed by centrifugation at 100,000 g or more. These extracts contain all the factors required for protein synthesis except ribosomes and mRNA and they are used to elongate and terminate the growing polypeptide chains attached to intact polyribosomes prepared from a variety of cells. The use of an S-100 system[41,42,43] is exactly the same as for a conventional *E. coli*, wheat germ or reticulocyte lysate extract, except that solutions of polyribosomes are added in place of purified mRNAs. As the polyribosomes are dissolved in a buffer containing K^+ and Mg^{2+}, the concentrations of these must be taken into account when making up the assay mixture.

Coupled Transcription/Translation Systems

Methods have been developed for the *in vitro* transcription of DNA by RNA polymerase, coupled with the translation of the synthesised RNA in the same extract.

The synthesis of active β-galactosidase in an *E. coli* lysate system primed with *E. coli* DNA has been described by Zubay *et al*.[44] A similar system has been used by Bottomley and Whitfield[45] to demonstrate the synthesis of chloroplast proteins in an *E. coli* lysate primed with chloroplast DNA.

DNA-directed synthesis of viral polypeptides in eukaryotic systems has been achieved by Roberts *et al*.,[46] using a wheat germ system supplemented with *E. coli* DNA-dependent RNA polymerase, and in a reticulocyte lysate system using endogenous DNA-dependent RNA polymerase, by Pelham.[47]

4. Assay for Protein Synthesis and Analysis of Products

Measurement of the Incorporation of Radioactivity into Protein

Protein synthesis is measured by determining the amount of radioactivity present in polypeptides after incubation of the *in vitro* translation mixture with mRNA and radioactive amino acid. The method for doing this is deceptively simple: proteins are insoluble in trichloroacetic acid (TCA) whereas unpolymerised labelled amino acid is soluble. However, radioactive amino acid molecules present as amino-acyl-tRNAs are also insoluble in TCA and must be removed. Furthermore, it is necessary to prevent non-specific-binding of soluble radioactive amino acid to the precipitates.

There are several ways of measuring the amount of acid-insoluble material. Probably the simplest method, especially when large numbers of samples have to be processed, is to dry a few microlitres of each sample onto a square or disc of Whatman filter paper and to batch-wash the filter papers. They should be marked with a pencil and washed successively in hot 5–10% TCA solution followed by cold TCA, ethanol, and finally ether.[48] The filter papers are then dried in an oven or with a hair dryer and counted in a scintillation counter using a toluene-based scintillation mixture. Samples containing [75]selenium should be counted in water with a γ-counter. Non-specific adsorption of radioactive amino acid to the filter papers is reduced by adding 0.5% unlabelled amino acid in the hot TCA wash. The hot TCA also releases radioactivity specifically bound to tRNA by hydrolysing the amino-acyl-tRNA. A modified washing procedure is required when using ^{35}S-cysteine, which reacts with free SH groups in proteins in the incubation mixture. The active groups can be blocked, after *in vitro* protein synthesis, by treatment with urea followed by carboxymethylation with iodoacetate by a modification of the method of Hirs,[49] described in a technical bulletin available from New England Nuclear Enterprises. Another procedure, which works quite well for [35]cysteine, is to add an equal volume of the following solution to the *in vitro* translation system *after* protein synthesis: 8M urea, 200 mM cysteine, 200 mM dithiothreitol. Aliquots of the mixture can then be dried on filter paper and subsequently washed with TCA, ethanol and ether as usual, before counting.

The products of *in vitro* protein synthesis are generally analysed to provide further information about the coding properties of the mRNA being studied. Specific protein products can be identified and isolated by using antibodies to purified proteins and, occasionally, specific translation products can be fractionated by virtue of their differential solubility in organic solvents. The products can also be fractionated by column chromatography. Generally, however, the *in vitro* translation mixture is treated in the presence of reducing agents (mercaptoethanol or dithiothreitol) and sodium dodecyl sulphate (SDS) and the proteins fractionated by polyacrylamide gel electrophoresis. Radioactive proteins are detected either in the wet gel or after drying the gel, by autoradiogrpahy or fluorography,[50,51] and the mobility compared with that of marker proteins. The γ-radiation from ^{75}Se-selenomethionine is rapidly detected by autoradiography. The β-radiation from ^{35}S takes less time to detect than that from ^{3}H or [14]C. With the latter isotopes, fluorography greatly speeds up detection on film. Identification of specific proteins can be accomplished by immuno-precipitation or fingerprinting (see Figure 7.6).

Optimising the Conditions

In vitro protein synthesis is very sensitive to pH, cation concentration, the presence of contaminants and, to a lesser extent, RNA concentration. Therefore, it is extremely important to use high-quality reagents and distilled water to make up buffers accurately and to optimise the reaction conditions.

The concentration of monovalent and divalent cations should be adjusted to the optimum concentration for translation of different mRNAs. The type of ion is also important. Higher rates of protein synthesis are obtained with K^+ than Na^+; Mg^{2+} is better than Ca^{2+}, which can inhibit translation; acetate as the counter ion gives better rates of protein synthesis than Cl^-. Detailed comparisons of the effect on *in vitro* protein synthesis of varying the concentration of different ions have been published for wheat germ[52,53] and for four different eukaryotic systems (see Figure 7.3).[54] It is important to test the ion concentration optima for translation of RNA fractions from different sources in order to ensure efficient and correct translation of the RNA.

The polyamines spermine and/or spermidine are frequently added to the wheat germ system. These increase protein synthesis, possibly by their interaction with tRNAs and amino-acyl tRNA synthetases.

Unlabelled amino acids are generally added to wheat germ *in vitro* translation mixtures. This is not necessary with the reticulocyte lysate. However, Amersham International now supply a fractionated and an unfractionated reticulocyte lysate. The former *is* depleted in amino acids and these are added back to the system before use. The wheat germ system obtained from Bethesda Research laboratories Inc. can be purchased with unlabelled amino acids added.

The optimum RNA concentration for protein snthesis varies with the type of RNA preparation (i.e. total polyribosomal RNA or purified poly(A)-containing RNA) and is different for each type of translation system. It is generally within the range 50–800 μg ml^{-1} (1–16 μg per 20 μl assay) (see Figure 7.4). The extracts should be incubated, with the

Figure 7.3: The Effect of Potassium Chloride and Potassium Acetate Concentrations on Protein Synthesis in vitro

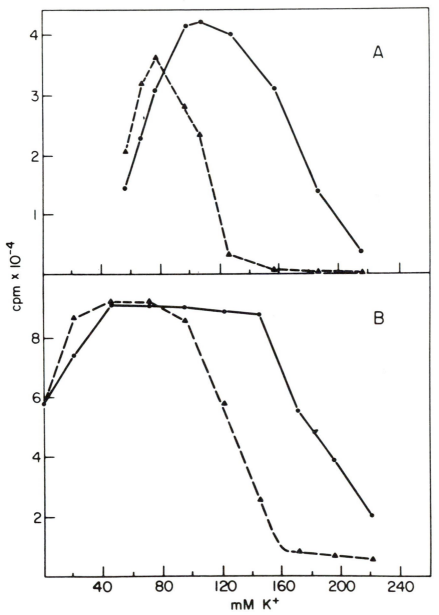

Notes: a. Wheat germ system with added rabbit globin mRNA.
   b. Rabbit reticulocyte lysate programmed by endogenous mRNA. Incorporation of ^{3}H-lysine into protein is expressed as cpm/5 µl aliquot. KCl (▲), K (oAc)(●).

Source: L.A. Weber, E.D. Hickey, P.A. Moroney and C. Baglioni, 'Inhibition of protein synthesis by Cl⁻', *J. Biol. Chem.*, 252 (1977).

Figure 7.4: The Effect of mRNA Concentrations on Protein Synthesis

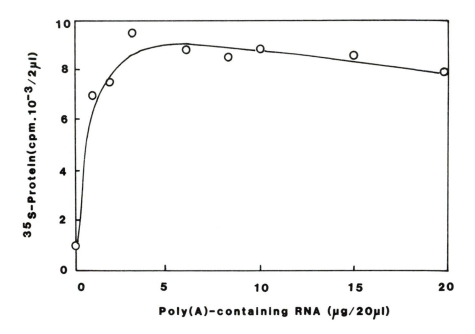

Notes: Poly(A)-containing RNA from french bean polyribosomes was used to direct the synthesis of protein in a wheat germ extract. Incorporation of ^{35}S-methionine into protein is expressed as cpm/2 μl aliquot. The amount of mRNA is shown per 20 μl incubation.

Source: A.B. Giles, 'Properties of plant polysomal messenger-RNA and its regulation by light'. Ph.D. thesis, University of Nottingham, 1976.

appropriate amount of mRNA, at 25 or 30°C for between 30 and 90 minutes (Figure 7.5).

The synthetic polynucleotide poly(U) has frequently been used as a pseudo-mRNA to demonstrate *in vitro* incorporation of amino acid (phenylalanine) into polypeptides and to optimise the conditions for protein synthesis. However, poly(U) does not contain an initiation triplet or ribosome-binding sequence and initiation of protein synthesis with this molecule does not resemble that of true mRNA. It is incorrect to extrapolate incorporation rates and/or optimum salt concentrations obtained with poly(U) to the *in vitro* translation of natural mRNAs. Wherever possible, natural mRNAs or a mixture of mRNAs, should be used to characterise the protein synthesising system and the quality of the products analysed by gel electrophoresis.

It is often convenient to test the activity of a translation system using a viral mRNA such as TMV RNA. This can either be prepared from virus-

Figure 7.5: Time Course for Translation of Different mRNAs in a Rabbit Reticulocyte Lysate

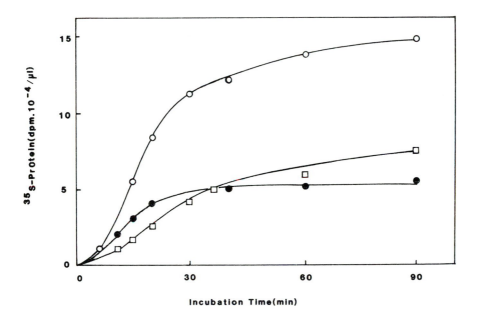

Note: Incorporation of ^{35}S-methionine into protein is expressed as cpm/μl of extract. Brome mosaic virus RNA (o), encephalo myocarditis virus RNA (□), mouse cell 3T6 virus RNA (o).

Source: Reproduced with permission from Amersham International plc, redrawn from Anon, 'Protein synthesis in cell-free systems', 1981.

infected tobacco plants[55] or purchased commercially. However, this only gives a general guide to the activity and optimum concentration for components in the extract. It is still necessary to optimise the conditions for the correct translation of the mRNAs one intends to study.

Precursor Proteins and Post-translational Modifications

Many proteins are initially synthesised as macromolecular precursors (pre-proteins), which undergo modification after synthesis. The two main types of modification are the removal of an amino-acid sequence (usually N-terminal) and the glycosylation of specific amino acid residues. The additional amino acid sequence in pre-proteins may function as a signal sequence to ensure the transport of proteins across membranes.[56,57,58,59]

In vivo, pre-proteins are rapidly 'processed' to their final size and composition and few copies of the unmodified molecules exist in the cell at

Figure 7.6: Identification of the Precursor to the Apoprotein of the Light-harvesting Chlorophyll a/b Protein of Barley

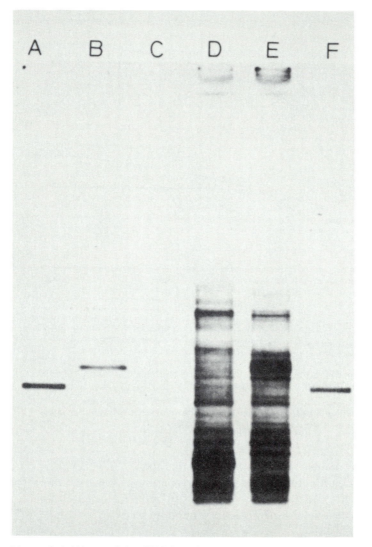

Note: Poly(A)-containing RNA from barley leaves was translated in a wheat germ system and ^{35}S-methionine incorporated into proteins. The products were fractionated by SDS-polyacrylamide gel electrophoresis and radioactive proteins detected by autoradiography. Tracks A and F show purified light-harvesting chlorophyll a/b protein. Tracks D and E show the translation products from dark-grown and illuminated barley leaves. Track B shows the immunoprecipitate from E, using an antibody to the light-harvesting chlorophyll a/b protein. No precipitation was detected with mRNA translation products from dark-grown plants (Track C).

Source: K. Apel and K. Kloppstech, 'The plastid membranes of barley (*Hordeum vulgare*). Light-induced appearance of mRNA coding for the apoprotein of the light-harvesting chlorophyll a/b protein', *Eur. J. Biochem. 85* (1978), 581–588.

any given time. In general, however, *in vitro* translation systems do not contain the necessary enzymes for processing, and the pre-proteins are stable and accumulate (Figure 7.6). They have a different electrophoretic mobility to that of the corresponding native protein, but can be identified by fingerprinting or by using specific antibodies.

Some progress has been made in developing *in vitro* translation systems which will process newly-synthesised proteins. Processing of some plant pre-proteins to the correct size has been achieved by the addition of membranes derived from microsomal preparations of dog pancreas or pea cotyledon.[60] However, the addition of membrane fractions in these experiments did not bring about processing of all the pre-proteins.

The most efficient and useful technique for *in vitro* processing studies is undoubtedly the *Xenopus* oocyte/egg system. Stimulation of the oocyte by micro-injection of mRNAs from a variety of organisms has been shown to result in synthesis of proteins which are correctly processed and packaged inside membrane vesicles.[61,62] Furthermore, proteins which, *in vivo*, are transported and secreted from cells, have been shown to be secreted from oocytes after injection of the appropriate mRNA.[63]

Further Reading

B. de Crombrugghe, 'DNA-dependent protein synthesis', in H. Weissbach and S. Pestka (eds.), *Molecular Mechanisms of Protein Biosynthesis* (Academic Press, New York and London, 1977), 603–625.

K. Moldave and L. Grossman (eds.), 'Nucleic Acids and Protein Synthesis', *Methods in Enzymology* (Academic Press, New York and London, 1974), 30F.

D.A. Shafritz, 'Messenger RNA and its Translation', in H. Weissbach and S. Pestka (eds.), *Molecular Mechanisms of Protein Biosynthesis* (Academic Press, New York and London, 1977), 555–601.

G. Zubay, '*In vitro* synthesis of proteins in microbial systems', *Ann. Rev. Genetics, 7* (1974), 267–287.

References

1. R.L.P. Adams, R.H. Burdon, A.M. Campbell, D.P. Leader and R.S. Smellie, '*The Biochemistry of the Nucleic Acids*', 9th edn (Chapman and Hall, London and New York, 1981).

2. P.R. Stewart and D.S. Letham (eds.), *The Ribonucleic Acids*, 2nd edn (Springer, New York, 1977).

3. D. Sagher, H. Grosfeld and M. Edelman, 'Large subunit ribulose-bisphosphate carboxylase messenger RNA from *Euglena* chloroplasts', *Proc. Nat. Acad. Sci., 73* (1976), 722–726.

4. S.T. Gibson, J. Speirs and C.J. Brady, 'On the response of plant ribosomes to ions and compatible organic solutes. II. *In vitro* synthesis on wheat ribosomes', submitted to *Plant, Cell and Environment*

5. A.J. Driesel, J. Speirs and H.J. Bohnert, 'Spinach chloroplast mRNA for a 32,000 dalton polypeptide. Size and localisation on the physical map of chloroplast DNA', *Biochim. Biophys. Acta, 610* (1980), 297–310.

6. J. Silverthorne and R.J. Ellis, 'Protein synthesis in chloroplasts. VIII. Differential

synthesis of chloroplast proteins during spinach leaf development', *Biochim. Biophys. Acta*, *607* (1980), 319–330.

7. R.H. Doi, *'Bacilus subtilis* protein synthesising system', in J.A. Last and A.I. Laskin (eds.), *Methods in Molecular Biology Vol. 1*; *Protein Biosynthesis in Bacterial Systems*, (Marcel Dekker Inc., New York, 1971), 67–88.

8. S.T. Bayley, 'Protein synthesis systems from halophytic bacteria', in J.A. Last and A.I. Laskin (eds.), *Methods in Molecular Biology Vol. 1*; *Protein Biosynthesis in Bacterial Systems* (Marcel Dekker Inc., New York, 1981), 89–110.

9. W. Bottomley and P.R. Whitfeld, 'Cell-free transcription and translation of total spinach chloroplast DNA', *Eur. J. Biochem.*, *93* (1979), 31–39.

10. G. Zubay, D.A. Chambers and L.C. Cheong, 'Cell-free studies on the regulation of the *lac* operon', in J. Beckwith and D. Zipser (eds.), *The Lactose Operon* (Cold Spring Harbor Laboratories, New York, 1970), 375–391.

11. Bottomley and Whitfeld, 'Cell-free transcription and translation'.

12. Ibid.

13. Zubay, Chambers and Cheong, 'Cell-free studies'.

14. J. Modolell, 'The S-30 system from *Escherichia coli*' in J.A. Last and A.I. Laskin (eds.), *Methods in Molecular Biology Vol. 1*; *Protein Biosynthesis in Bacterial Systems*, (Marcel Dekker Inc., New York, 1971), 1–65.

15. J. Speirs and D. Grierson, 'Isolation and Characterisation of 14S-RNA from spinach chloroplasts', *Biochim. Biophys. Acta*, *521* (1978), 619–633.

16. H.R.B. Pelham and R.J. Jackson, 'An efficient mRNA-dependent translation system from reticulocyte lysates', *Eur. J. Biochem.*, *67* (1976), 247–256.

17. B.E. Roberts and B.M. Patterson, 'Efficient translation of tobacco mosaic virus RNA and rabbit globin 9S RNA in a cell-free system from commercial wheat germ', *Proc. Nat. Acad. Sci.*, *70* (1973), 2330–2334.

18. K. Marcu and B. Dudock, 'Characterisation of a highly efficient protein synthesising system derived from commercial wheat germ', *Nucleic Acids Res.*, *1* (1974), 1385–1397.

19. Pelham and Jackson, 'An efficient mRNA-dependent translation system'.

20. T. Hunt and R.J. Jackson, 'The rabbit reticulocyte lysate as a system for studying mRNA' in R. Neth, R.C. Gallo, S. Spiegelman and F. Stohlman (eds.), *Modern Trends in Human Leukaemia* (J.F. Lehmans Verlag, Munich, 1974), 300–307.

21. M.B. Matthews and A. Korner, 'Mammalian cell-free protein synthesis directed by viral ribonucleic acid', *Eur. J. Biochem.*, *17* (1970), 328–338.

22. W. Zagorski, 'Preparation and characteristics of a wheat embryo cell-free extract active in the synthesis of a high molecular weight viral polypeptide', *Anal. Biochem.*, *87* (1978), 313–333.

23. M.H. Tuite, J. Plesset, K. Moldave and C.S. McLaughlin, 'Faithful and efficient translation of homologous and heterologous mRNAs in an mRNA-dependent cell-free system from *Saccharomyces cerevisiae*', *J. Biol. Chem.*, *255* (1980), 8761–8766.

24. A.R. Carlier and W.J. Peumans, 'The rye embryo system as an alternative to the wheat germ system for protein synthesis *in vitro*', *Biochim. Biophys. Acta*, *447* (1976), 436–444.

25. A.R. Carlier, W.J. Peumans and A. Manickam, 'Cell-free translation in extracts from dry and germinated mung bean primary axes', *Plant Sci. Letters*, *11* (1978), 207–216.

26. W.J. Peumans, A.R. Carlier and B.M. Delaey, 'Preparation and characterization of a highly active cell-free protein-synthesizing system from dry pea primary axes', *Plant Physiol.*, *66* (1980), 584–587.

27. I. Fischer and K. Moldave, 'Preparation and characterization of a cell-free system from Chinese hamster ovary cells that translates natural messenger ribonucleic acid and analysis of intermediary reactions', *Anal. Biochem.*, *113* (1981), 13–26.

28. Roberts and Patterson, 'Efficient translation of tobacco mosaic virus'.

29. Marcu and Dudock, 'Characterisation of a protein synthesising system'.

30. A. Marcus, D. Effron and D.P. Weeks, 'The wheat embryo cell-free system', in S.P. Colowick and N.O. Kaplan (eds.), *Methods in Enzymology* (Academic Press, New York and London, 1974), 749–54.

31. Marcu and Dudock, 'Characterisation of a protein synthesising system'.

32. R.T. Schimke, R.E. Rhoads and S. McKnight, 'Assay of ovalbumin mRNA in reticulocyte lysate', *Methods in Enzymology*, *30* (1974), 694–701.

33. Hunt and Jackson, 'The rabbit reticulocyte lysate'.

34. C.D. Lane, G. Marbaix and J.B. Gurdon, 'Rabbit haemoglobin synthesis in frog cells: the translation of reticulocyte 9S RNA in frog oocytes', *J. Mol. Biol.*, *61* (1971), 73–91.
35. J.B. Gurdon, C.D. Lane, H.R. Woodland and G. Marbaix, 'Use of frogs eggs and oocytes for the study of messenger RNA and its translations in living cells', *Nature*, *233* (1971), 177–82.
36. D.L. Stephens, T.J. Miller, L. Silver, D. Zipser and J.E. Mertz, 'Easy to use equipment for the accurate microinjection of nanolitre volumes into the nuclei of amphibian oocytes', *Anal. Biochem.*, *114* (1981), 299–309.
37. T.J.V. Higgins and D. Spencer (1981), 'Precursor forms of pea vicilin subunits. Modification by microsomal membranes during cell-free translation', *Plant Physiol.*, *67* (1981), 205–11.
38. B.A. Larkins, K. Pedersen, A.K. Handa, W.J. Hurkman and L.D. Smith, 'Synthesis and processing of maize storage proteins in *Xenopus laevis* oocytes', *Proc. Nat. Acad. Sci.*, *76* (1979), 6448–52.
39. C.D. Lane, A. Coleman, T. Mohun, J. Morser, J. Champion, I. Koundes, R. Craig, S. Higgins, T.C. James, S.W. Applebaum, R.I. Ohlsson, E. Paucha, M. Houghton, J. Mathews and B.J. Miflin, 'The *Xenopus* oocyte as a surrogate secreting system. The specificity of protein export', *Eur. J. Biochem.*, *111* (1980), 225–35.
40. Ibid.
41. P. Musk, A.K. Chakravorty and K.J. Scott, 'Protein synthesis by polysomes isolated from wheat leaves', *Plant and Cell Physiol.*, *20* (1979), 1359–69.
42. S. Fikuski, K. Ishikawa and K. Sasaki, '*In vitro* protein synthesis during germination and vernalization in winter wheat embryos', *Plant and Cell Physiol.*, *18* (1977), 969–77.
43. L. Rhyanen, P.N. Graves, G.M. Bressan and D.J. Prockop, 'Synthesis of an elastin component of molecular weight about 70,000 by polysomes from chick embryo aortas', *Arch. Biochem. Biophys.*, *185* (1978), 344–51.
44. Zubay, Chambers and Cheong, 'Cell-free studies'.
45. Bottomley and Whitfield, 'Cell-free transcription and translation'.
46. B.E. Roberts, M. Gorecki, R.C. Mulligan, K.J. Danna, S. Rozenblatt and A. Rich, 'Simian virus 40 DNA directs synthesis of authentic viral polypeptides in a linked transcription-translation cell-free system', *Proc. Nat. Acad. Sci.*, *72* (1975), 1922–6.
47. H.R.B. Pelham, 'Use of coupled transcription and translation to study mRNA production by vaccinia cores', *Nature*, *269* (1977), 532–4.
48. F.J. Bollom, 'Filter paper disk techniques for assaying radioactive molecules', *Methods in Enzymology*, *12B* (1968), 169–173.
49. C.H.W. Hirs, 'Reduction and S-carboxymethylation of proteins', *Methods in Enzymology*, *11* (1967), 199–203.
50. W.M. Bonner and R.A. Laskey, 'A film detection method for tritium-labelled proteins and nucleic acids in polyacrylamide gels', *Eur. J. Biochem.*, *46* (1975), 83–88.
51. R.A. Laskey and A.D. Mills, 'Quantitative film detection of ^{3}H and ^{14}C in polyacrylamide gels by fluography', *Eur. J. Biochem.*, *56* (1975), 335–341.
52. R.G. Wyn Jones, C.J. Brady and J. Speirs, 'Ionic and osmotic relations in plant cells' in D.L. Laidman and R.G. Wyn Hones (eds.), *Recent Advances in the Biochemistry of Cereals* (Academic Press, London and New York, 1979), 63–103.
53. Gibson, Speirs and Brady, 'On the response of plant ribosomes'.
54. L.A. Weber, E.D. Hickey, P.A. Moroney and C. Baglioni, 'Inhibition of protein synthesis by Cl^-', *J. Biol. Chem.*, *252* (1977), 4007–4010.
55. Marcus, Effron and Weeks, 'The wheat embryo cell-free system'.
56. G. Blobel and B. Dobberstein, 'Transfer of proteins across membranes. I. Presence of proteolytically processed and unprocessed nascent immunological light chains on membrane-bound ribosomes of murine myeloma', *J. Cell. Biol.*, *67* (1975), 835–851.
57. G. Blobel and B. Dobberstein, 'Transfer of proteins across membranes. II. Reconstitution of functional rough microsomes from heterologous components', *J. Cell Biol.*, *67* (1975), 852–62.
58. V.R. Lingappa, J.R. Lingappa and G. Blobel, 'Chicken ovalbumin contains an internal signal sequence', *Nature*, *281* (1979), 117–21.
59. P.E. Highfield and R.J. Ellis, 'Synthesis and transport of the small subunit of chloroplast ribulose bisphosphate carboxylase', *Nature*, *271* (1978), 420–4.
60. Higgins and Spencer, 'Precursor forms'.

61. Larkins, Pedersen, Handa, Hurkman and Smith, 'Synthesis and processing'.
62. Lane, Coleman, Mohun, Morser, Champion, Koundes, Craig, Higgins, James, Applebaum, Ohlsson, Paucha, Houghton, Mathews and Miflin, 'The *Xenopus* oocyte'.
63. Ibid.

8 LABELLING DNA *IN VITRO* – NICK TRANSLATION

Christopher G.P. Mathew

CONTENTS

1. Introduction

If a single-stranded DNA molecule is placed in contact with a complementary single-stranded DNA sequence, the two molecules will associate with one another by hydrogen bonding between the bases on their respective strands. This association, or hybridisation, forms the basis of very powerful techniques for detecting and quantifying specific nucleic acid sequences. Whether the hybridisation is done in solution or to nucleic acid immobilised on filters, a radioactively labelled probe will be required. The enzyme DNA polymerase I from *E. coli* catalyses a reaction which can be used to replace existing unlabelled nucleotides in DNA with radioactive ones. The reaction has been called nick translation, and is very widely used in molecular biology.[1,2] It has nothing whatever to do with the translation of messenger RNA into a polypeptide chain on the ribosomes.

2. The Reaction Mechanism

The mechanism of the reaction is outlined in diagrammatic form in Figure 8.1. The initial step in the procedure is to create free 3'-hydroxyl groups within the unlabelled DNA ('nicks') by means of a nuclease such as pancreatic deoxyribonuclease. The DNA polymerase I will then catalyse the addition of a nucleotide residue to the 3'-hydroxyl terminus of the nick. At the same time, the 5' to 3' exonuclease activity of this enzyme will eliminate the nucleotide unit from the 5'-phosphoryl terminus of the nick. The net result of the reaction will be the incorporation of a new nucleotide with a free 3'-OH group at the position where the original nucleotide was excised. The nick will, therefore, have been shifted along by one nucleotide unit in a 3' direction. This 3' shift, or translation, of the nick will result in sequential addition of new nucleotides to the DNA while the pre-existing nucleotides will be removed. The DNA thus acts as both primer for the DNA polymerase, because of its free 3'-OH group, and as template, because the opposite strand of the DNA duplex dictates what type of nucleotide is to be incorporated. If radioactively labelled deoxyribonucleoside triphosphates (dNTPs) are used as substrates, the original unlabelled nucleotides in the DNA will be replaced by labelled ones. By means of this 'hot for cold swop' of nucleotides, about 50% of the residues in the DNA can be labelled.[2]

161

Figure 8.1: Schematic Representation of the Nick Translation Reaction

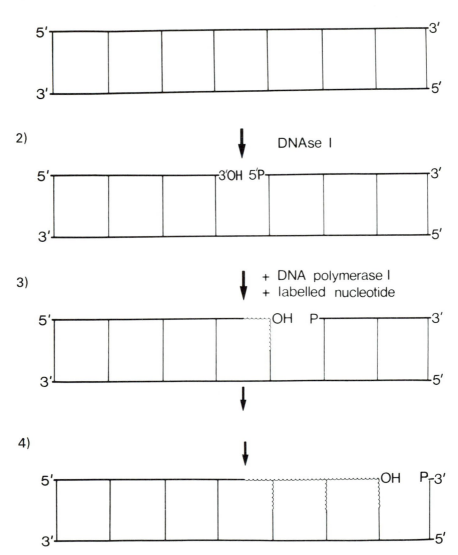

1. Intact double-stranded DNA molecule.
2. The DNAse I has produced a nick with a 3'-hydroxyl terminus in the DNA duplex.
3. The first nucleotide on the 5'-phosphate side of the nick has been replaced with a labelled nucleotide. This now has a 3'-OH terminus, so the nick has been translated one position along the chain in a 5' to 3' direction.
4. The nick has been translated two more positions along the DNA.

3. Experimental

Procedure

Details of the experimental procedure, buffer compositions etc. can be found in Rigby *et al.*[2] and Favaloro *et al.*[3] The composition of the reaction mixture is as follows:

(i)   The DNA to be labelled
(ii) The four deoxyribonucleoside triphosphates, dCTP, dATP, dGTP and dTTP, one or more of which will be labelled with ^{32}P or ^{3}H.
(iii) Deoxyribonuclease I (optional – see below)
(iv) *E. coli* DNA polymerase I

The reaction is usually carried out at about 15°C for one to two hours and is terminated by the addition of EDTA. Unincorporated nucleotides are separated from the DNA by chromatography on Sephadex G-50, and the specific activity of the DNA determined by counting the radioactivity in the excluded peak. The DNA can then be purified by phenol extraction.

Properties of Labelled DNA

The most important property of the DNA probe is its specific activity. The higher the specific activity, the more sensitive and accurate the detection and quantification of specific sequences will be. In order to make a probe of high specific activity, sufficient labelled nucleotide must be added to the reaction mixture to replace all the corresponding nucleotides in the DNA. For instance, if 4,000 pmol of DNA are to be labelled, then 1,000 pmol of say dCTP will be required to replace all the 'cold' dCTP, whereas 100 pmol could only replace 10% of it. In practice, an incorporation of 30–50% is usually obtained. The specific activity of the probe will obviously also depend on the specific activity of the nucleotide(s) used in the nick translation. dNTPs with activities of 3,000 Curies/mmol are commercially available, and use of only one labelled nucleotide at this activity can be expected to produce a probe with a specific activity of $1-3 \times 10^8$ cpm/µg. Adequate detection of unique sequences by Southern blotting, for example, requires a probe with an activity of 10^8 cpm/µg.

The amount of deoxyribonuclease I added to the reaction is also an important parameter. If the DNAse concentration is high, many nicks will be generated, and the DNA polymerase I can begin translating at many sites. The reaction rate will therefore be high, and a probe of high specific activity will be generated in the standard incubation time. However the single-stranded length of the labelled probe, and hence its rate of hybridisation to complementary sequences, will be reduced at high DNAse concentrations. For most applications, sufficient DNAse is added to produce a probe with a single-stranded length of about 400 nucleotides. An intact probe of lower specific activity may be produced if DNAse is left out of the reaction altogether.

Another important requirement for a probe is that it should be labelled throughout at a uniform specific activity. If, for example, a section of the probe is poorly labelled, a restriction fragment corresponding to that part of the probe might not be detected in a Southern blotting experiment. Rigby *et al.*[2] have shown that probes are uniformly labelled by nick translation, so DNAse I nicks DNA randomly and DNA polymerase I translates each nick with equal probability.

The stability of a nick translated probe on storage is dependent on the type of isotopic label used.[4] A ^{32}P labelled probe with a specific activity of 10^8 cpm/μg is stable for seven to eight days, after which its single-stranded length declines rapidly. ^{3}H probes, however, may be stable for six to nine months.

Alternative Methods for Labelling DNA

DNA can be labelled *in vivo* by supplying radioactive precursors to a tissue culture system (method described by Favaloro *et al.*[3]). For example, labelled viral DNA can be extracted from virus-infected cells which have been grown in the presence of ^{32}P. Nick translation has several advantages over this approach: labelled DNA can be produced even when no appropriate tissue culture system is available; DNA samples can be purified and stored unlabelled, then labelled rapidly when required; DNA can be labelled to a much higher specific activity by nick translation.

Another possible approach is 3'-end labelling of DNA with T4 DNA polymerase.[5] In the absence of free nucleotides, the 3' exonuclease activity of this enzyme will act on the 3' ends of double-stranded DNA to produce DNA fragments with recessed 3' termini. Upon addition of labelled nucleotides, the polymerase activity of the enzyme fills in the ends. Probes of very high specific activity can be produced by this method, but the DNA will not be uniformly labelled. Thus any fragmentation of a probe labelled by this method could generate pieces of DNA with little or no label.

4. Applications

Nick translated probes are widely used in molecular biology for detecting and quantifying specific sequences. For example:

(i) Southern blotting – nick translated probes are routinely used for the detection of specific sequences on Southern blots (see Chapter 15).
(ii) Quantitative analysis of specific sequences – this is a solution hybridisation method whereby the concentration of DNA sequences homologous to a labelled probe can be measured in the presence of a great excess of unrelated DNA. The basis of the assay is kinetic, i.e. the time

taken for a complementary sequence to reassociate with the probe reflects the concentration of that sequence. Rigby *et al.*[2] used nick translated ^{32}P-SV40 DNA to detect one SV40 DNA molecule per haploid mouse genome in less than 100 μg of mouse cellular DNA.

(iii) Screening for recombinant DNA molecules – nick translated probes are used to identify bacterial colonies which contain recombinant plasmids or phages (see Chapter 12).

(iv) Chromosomal location of cloned DNA – the technique of *in situ* hybridisation enables one to establish on which chromosome a particular nucleotide sequence resides. Chromosomes are squashed onto glass microscope slides, denatured, and incubated with nick translated probes for specific sequences. Unhybridised material is then washed off, and the position of hybridisation localised by autoradiography.[6]

(v) Labelling active chromatin – chromatin that is transcriptionally active is known to be preferentially degraded by DNAse I. Brief treatment of nuclei with DNAse I, followed by DNA polymerase I, therefore results in selective incorporation of labelled nucleotides into transcribed DNA.[7] Radioactive 'tagging' of the active chromatin fraction is a useful tool in studies of gene expression.

(vi) Visualisation of restriction fragments – if one wishes to construct a restriction map of a DNA of which very limited quantities are available, the DNA can be nick translated (in the absence of DNAseI) and an autoradiograph made of the gel. Whereas a conventional ethidium bromide stain will only detect about 100 ng of DNA per band, a DNA which has been labelled to a specific activity of 10^6–10^7 cpm/μg can be detected at the picogram level.

Further Reading

The Radiochemical Centre, Amersham, 'Labelling of DNA with ^{32}P by nick translation', Technical bulletin 80/3, (1980).

References

1. T. Maniatis, A. Jeffrey and D.G. Kleid, 'Nucleotide sequence of the rightward operator of phage λ', *Proc. Natl. Acad. Sci. USA.*, *72* (1975), 1184–8.
2. P.W.J. Rigby, M. Dieckmann, C. Rhodes and P. Berg, 'Labelling DNA to high specific activity in vitro by nick translation with DNA polymerase I', *J. Mol. Biol.*, *113* (1977), 237–51.
3. J. Favaloro, R. Treisman and R. Kamen, 'Transcription maps of polyoma virus-specific RNA : Analysis by two-dimensional nuclease SI gel mapping', in L. Grossman and K. Moldave (eds.), *Methods in Enzymology*, *65* (1980), part 1, 718–49.
4. J.K. Mackey, K.H. Brackmann, M.R. Green and M. Green, 'Preparation and characterization of highly radioactive in vitro labelled adenovirus DNA and DNA restriction fragments', *Biochemistry*, *16* (1977) , 4478–84.
5. P.H. O'Farrel, E. Kutter, and M. Nakanishi, 'Replacement synthesis method of labelling DNA fragments', *Molec. Gen. Genetics*, *179* (1980), 421–35.

6.   F.H. Ruddle, 'A new era in mammalian gene mapping : somatic cell genetics and recombinant DNA methodologies', *Nature*, *294* (1981), 115–20.
7.   A. Levitt, R. Axel and H. Cedar, 'Labelling active chromatin by nick translation', *Develop. Biol.*, *69* (1979), 496–501.

9 SYNTHESIS OF cDNA FOR MOLECULAR CLONING

Brian G. Forde

CONTENTS

1. Introduction

Within the last decade the newly developed techniques of genetic manipulation have enabled us to isolate, in purified form, milligram quantities of a single eukaryotic gene. As a result of the isolation of genes such as those for globin, insulin and the immunoglobulins, a series of fundamental advances have been made in our understanding of the organisation of the eukaryotic genome. The presence of sequences which interrupt the structural gene sequences (introns), for example, was unsuspected prior to the analysis of recombinant eukaryotic DNA.

The foundations for these advances were laid in the early 1970s by the discovery of enzymes (restriction endonucleases) which could cut DNA at specific sites, and other enzymes (ligases) which allowed DNA molecules from different sources to be joined. Also crucial was the development, at about the same time, of a way of introducing exogenous DNA into *Escherichia coli* (see Chapter 12). By combining these techniques it was possible to insert a foreign DNA fragment into a bacterial 'vector' (plasmid or bacteriophage) and then to amplify the hybrid molecule by propagating it in a line of genetically identical cells (a clone). If a mixture of hybrid molecules containing different DNA inserts is generated, then an individual clone will harbour only one kind of hybrid molecule. Once the clone which carries the DNA fragment of interest has been identified, a bulk preparation of the cells can be made and the hybrid plasmid or phage DNA isolated.

The enormous size of the eukaryotic genome, however, poses major problems for the identification of cloned DNA sequences. The haploid genome of the rabbit, for example, contains around 3×10^9 nucleotide pairs of DNA. To make reasonably certain of cloning a single-copy gene it is therefore necessary to generate over 200,000 clones each carrying about 17×10^3 nucleotide pairs of rabbit DNA. To screen so many clones, without an enormous expenditure of time and effort, requires the use of a highly specific hybridisation probe; usually this is provided by a cloned complementary DNA sequence. Complementary DNA (cDNA) is the single-stranded product of the reverse transcription of messenger RNA (mRNA). In comparison to nuclear DNA, the mRNA population in a eukaryotic cell is highly enriched for structural gene sequences. This is because only a small fraction of the genome is expressed as mRNA in the cytoplasm: in an extreme case, such as that of globin in mammalian reticulocytes, almost all the mRNA may code for a single type of protein. As a result, the number of clones which must be constructed and screened to obtain a DNA sequence in the form of cDNA is usually at least 100-fold less than the number of clones needed to make a 'library' of genomic DNA fragments. Of course, cDNA clones will lack both introns and the sequences which flank the transcribed region of the structural gene. To obtain clones carrying these sequences it is therefore still necessary to construct and screen a genomic library (see also Section 7).

The procedure for the preparation of cDNA for cloning can be divided into six stages (Figure 9.1). The first four of these concern the synthesis and double-stranding of the cDNA and are the subject of this chapter. The remaining steps, involving techniques for the insertion of the double-stranded cDNA (ds-cDNA) into a plasmid, are discussed in Chapter 10.

2. Outline of the Method

Three enzymatic reactions form the basis for the *in vitro* conversion of mRNA sequences into ds-cDNA (Figure 9.1). The experimental conditions for each have been reviewed[1,2] and detailed protocols for the entire procedure have been published.[3,4]

Reverse Transcription

In 1970 two laboratories simultaneously reported the isolation of a viral enzyme which was capable of synthesising DNA on an RNA template.[5,6] Within two years the enzyme, 'reverse transcriptase' had been successfully used by several laboratories to synthesize a cDNA copy of globin mRNA (reviewed by Efstratiadis and Villa-Komaroff).[7] The conditions which were worked out for the reverse transcription of globin mRNA have subsequently been shown to be applicable, with minor modifications, for every mRNA so far tested.

The reverse transcriptase widely used for cDNA synthesis is isolated from cells infected with avian myeloblastosis virus (AMV). To begin transcription of its template, the reverse transcriptase requires an oligonucleotide primer with a free 3'-OH group. The primer is normally provided by a short oligo(dT) molecule, 12–18 nucleotides in length. The oligo(dT) chain is able to anneal to the sequence of polyadenylic acid (poly(A)) residues which is conveniently located at the 3' end of most eukaryotic mRNAs.

The immediate product of reverse transcription is an mRNA:cDNA hybrid. Before proceeding, the mRNA is removed by treatment of the hybrid with alkali, which hydrolyzes the RNA but not the DNA.

Many of the cDNA transcripts obtained are often found to be shorter than their mRNA templates. This can be explained in a number of ways, as discussed later (Section 3), but the net result is that the single-stranded cDNA is generally heterogeneous in size, each molecule having an oligo(dT) sequence of varying length at its 5' terminus.

Figure 9.1: Diagram showing Six Stages in the Synthesis of Double-stranded cDNA and its Insertion into a Plasmid Vector for Cloning

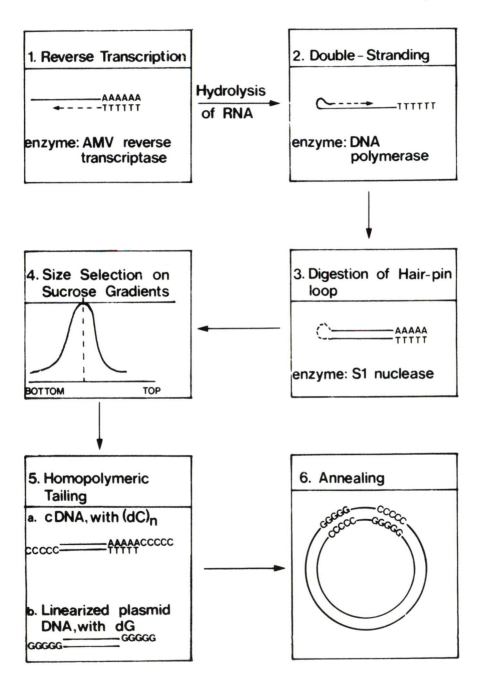

Double-stranding

Before it can be cloned the cDNA must be converted into a double-stranded molecule. Like the first reaction the second reaction depends on a primer. However, in this case DNA synthesis is primed by a hairpin loop which is formed at the 3' end of both full-length and partial reverse transcripts. The self-primed copying of the first strand is usually catalysed by *E. coli* DNA polymerase I, although the Klenow fragment of this (lacking the 5'-exonuclease activity), reverse transcriptase and T4 DNA polymerase have also been used.[8]

S1 Nuclease Digestion

Use of the hairpin loop as the primer for double-stranding means that the second strand is covalently linked to the first. To sever this link the DNA is treated with S1 nuclease, an enzyme from *Aspergillus oryzae* which, under appropriate conditions, specifically digests single-stranded DNA. Since the hairpin loop is partially single-stranded it is removed by this treatment, as is any cDNA which has remained uncopied in the second reaction.

Size-fractionation

After S1 nuclease digestion there will usually be a large number of incomplete ds-cDNA molecules, in addition to some which represent entire mRNAs. Some of these incomplete ds-cDNAs will lack sequences representing the 5' terminal region of the mRNA (due to incomplete first-strand synthesis), others will lack sequences from the 3' terminal region of the mRNA (due to incomplete copying of the first strand) and a third group will represent only the central portion of the mRNA (due to incomplete copying of an incomplete first strand). As a result, even those ds-cDNA molecules derived from a single species of mRNA will be heterogeneous is size. Since smaller polynucleotides are cloned much more readily than larger ones it is important to remove the low molecular weight ds-cDNA before proceeding. For this reason the ds-cDNA is usually fractionated by centrifugation through a sucrose gradient[9] and the slower sedimenting material discarded.

3. Optimising for Size and Yield

A large number of factors are known to influence the efficiency with which mRNA sequences are converted into full-length ds-cDNA. These can be considered under four headings.

The RNA Template

Not all mRNAs are transcribed equally efficiently by AMV reverse transcriptase. When four mRNAs from the hen oviduct were copied with reverse transcriptase it was found that one, conalbumin mRNA, produced a high proportion of incomplete transcripts.[10] Later studies showed that the problems with conalbumin mRNA could be partly overcome by denaturing the mRNA with methylmercury hydroxide.[11] Therefore it appears that, in this case at least, the inefficiency of the mRNA as a template for reverse transcriptase was due to regions of the RNA being extensively base-paired. It is easy to imagine that such areas of secondary structure might interfere with progress of the enzyme along the mRNA.

Reagent and Enzyme Concentrations

Several detailed studies have been made of the optimum conditions of reverse transcription.[12] However there has been little general agreement about the most suitable conditions, presumably because of differences between the mRNA templates used and in the purity of the reverse transcriptase preparations.[13] To ensure maximum yields of full-length cDNA it is therefore necessary to establish optimum conditions for each mRNA template and for each new batch of reverse transcriptase. The most important parameters to be checked are the concentrations of the monovalent and divalent cations and of the enzyme and its template. There does seem to be agreement that the total concentration of the substrates, the deoxynucleotide triphosphates, should be high[14] (greater than 3 mM). A sequence of steps for optimisation has been suggested by Maniatis *et al.*,[15] who also point out the necessity of taking into account the contribution of the enzyme storage buffer to the pH and monovalent cation concentration.

For the second reaction, catalyzed by DNA polymerase 1, the nature of the original mRNA template seems to be of less importance than in the first reaction.[16] The critical parameters are the temperature, which must be kept low (15°C), the time of incubation (at least four hours), the nature and concentration of the monovalent cation, the enzyme concentration (which should be saturating for the concentration of cDNA in the reaction) and a pH which suppresses the nucleolytic activities of DNA polymerase 1.[17] In one of the few detailed optimisation studies to have been carried out on the double-stranding reaction, Wickens *et al.* found that different preparations of DNA polymerase I required different incubation conditions to give full-length second strands.[18] Therefore, although it should not be necessary to use different reaction conditions for different cDNAs, it does appear to be important to optimise the conditions for each batch of the enzyme.

The digestion of ds-cDNA with S1 nuclease should be performed at high salt concentration (0.2–0.3 M) to prevent the nicking of double-stranded DNA.[19] Although the enzyme concentration used is always in excess of that required, it is nevertheless advisable to titrate each S1 nuclease preparation to ensure that the concentration used is one to which double-stranded DNA is resistant.[20]

Ribonuclease Contamination

A factor of major importance in reducing the size and yield of reverse transcripts is the presence of contaminating ribonucleases which degrade the mRNA template. For this reason all solutions used for the first reaction (as well as for subsequent steps) should be autoclaved before use. Similarly all glass and plastic reaction vessels or chromatography columns should be sterilised. The mRNA must be extensively purified, usually including one or two passages over an oligo(dT) cellulose column (see Chapter 6). The source of ribonuclease activity which causes the greatest problem is often the ribonuclease H which is associated with the reverse transcriptase preparation itself. The seriousness of this contamination will, of course, vary from one enzyme preparation to another and may at least partly be overcome by further purification.[21]

Sodium pyrophosphate has been found to improve the size of reverse transcripts,[22] presumably through its ability to inhibit ribonuclease activity. However, more recently a ribonuclease inhibitor extracted from human placenta, and now available commercially, has given promising results in improving the yield of full-length transcripts of the very large bovine and human thyroglobulin mRNAs.[23,24] It is interesting to note that Buell and his colleagues found that once they had eliminated the problem of nucleases, the proportion of full-length transcripts obtained was not significantly affected by any of the other reaction conditions they investigated.[25] They therefore concluded that some of the conditions previously reported to increase the proportion of full-length cDNA were not operating on reverse transcriptase itself, but were simply inhibiting the associated ribonuclease activity.

Losses Occurring during Manipulations

The amounts of cDNA synthesized are usually very small (< 1 µg) and care must therefore be taken to reduce losses during the succeeding series of manipulations. Between each enzyme reaction the sample is normally extracted with phenol and chromatographed over Sephadex before being concentrated by ethanol precipitation.[26,27] Losses during these procedures can be minimised by using only solutions which have been membrane-filtered before autoclaving, and by using plastic disposable tubes and syringes in preference to glass whenever possible. Small (1.5 ml) polypropylene centrifuge tubes, in which the reactions are usually carried out, and any glassware with which the cDNA comes in contact, should be siliconised by treatment with dimethyldichlorsilane. Siliconisation prevents adsorption of the DNA to the walls of the tubes, which is a particularly serious problem with glass. Radioactive labelling of the cDNA, by including a ^{32}P-labelled deoxynucleotide triphosphate in the first reaction, makes it much easier to follow its recovery from chromatography columns and ethanol precipitations.

So that the number of manipulations are reduced to a minimum, a method has been devised to perform the first and second reactions in the same tube, omitting the intermediate purification procedures.[28] The first reaction mixture is simply boiled for three minutes to free the cDNA of its mRNA template and the components necessary for the second reaction are added.

Regardless of the precautions taken, the final yields of ds-cDNA are usually no more than a few percent of the mass of the original RNA template. Therefore, since 20–100 ng of ds-cDNA is usually considered to be the minimum requirement for cloning, it is advisable to begin the synthesis of cDNA with at least 2–5 μg of poly A$^+$ RNA.

4. Monitoring the Reactions and their Optimisation

The progress and results of the enzyme reactions can be monitored in two principal ways: for yield, by measurement of the incorporation of radioactive deoxynucleotides, and for size, by electrophoretic analysis of the radioactively labelled DNA in denaturing acrylamide or agarose gels.

The incorporation of ^{32}P- or ^{3}H-labelled precursors is estimated by removing aliquots of the reaction mix during and after the reaction. Radioactivity which is precipitable with trichloracetic acid is then assayed in a liquid scintillation counter (see Goodman and MacDonald[29] for a description of the procedure). Since the specific activity of the radioactive precursor is known, this gives a means of estimating the yield of the reaction product.

To determine the size of the products obtained, a sample is fractionated in one of two alternative electrophoretic systems. The system chosen will largely depend on the expected size of the product: for larger DNA molecules, over 1000 nucleotides in length, agarose gels containing alkali can be used;[30] for DNA molecules with chain lengths of less than 1000 nucleotides a 5% polyacrylamide gel containing 98% formamide is preferable (see Chapter 14).[31] Reaction products labelled with ^{32}P are detected by covering the wet gel with cling-film and autoradiographing. The molecular weights of the DNA molecules can be accurately determined by running DNA size markers in the same gel.[32] An example of the results obtained when a heterogeneous population of mRNA from barley endosperm was reverse transcribed is seen in Figure 9.2 (lanes 2–5). The autoradiograph shows that the cDNA, as expected, is heterodisperse, but that a number of distinct bands are visible. In this instance, because of the heterogeneity of the mRNA template, it is not clear whether these bands represent full-length transcripts of small mRNAs, or whether they resulted from the termination of transcription at defined sites on longer mRNAs.

Optimisation of reverse transcription will usually begin by establishing the conditions which give the best yield of cDNA per μg mRNA, as

Figure 9.2: Autoradiograph of ^{32}P-labelled cDNA which was Synthesized from Barley Endosperm Poly A$^+$ RNA

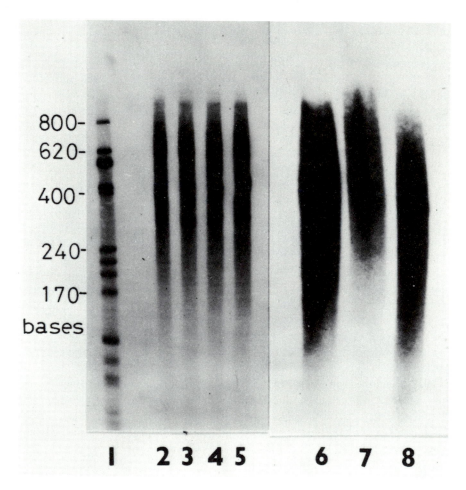

Note: The reaction products of reverse transcription and double-stranding were analysed on 5% polyacrylamide gels containing 98% formamide. [^{32}P]dATP was included in the first reaction but not in the double-stranding, so that only the first (cDNA) strands were detected on the autoradiograph of the dried-down gel. Track 1: *Hae*III fragments of pMB9 which had been ^{32}P-labelled by 'nick translation' to provide size markers; Tracks 2–5: products of reverse transcription of polyadenylated RNA from membrane-bound polysomes of barley endosperm. Samples were taken for electrophoresis after 10, 30, 60 and 90 minutes respectively. In this experiment there was little further reverse transcription after the first 10 minutes, perhaps because of degradation of the template by ribonucleases. Track 6: products of reverse transcription obtained in another experiment; Track 7: products obtained after double-stranding of the cDNA; Track 8: the double-stranded cDNA after S1 nuclease digestion. The reaction conditions used for reverse transcription, double-stranding and S1 nuclease digestion have been described.[33]

measured by the incorporation of radioactive label. A cDNA yield of 20–35% of the mass of the mRNA can usually be achieved. If the problem of ribonuclease has been overcome these should also be the conditions which give the longest transcripts and this can be checked by electrophoresis. However, any problems which are peculiar to the mRNA species of interest, such as secondary structure, will be difficult to detect and solve if the mRNA is only available in an impure form.

If the cDNA has been labelled with ^{32}P, the second reaction can be followed by supplying a ^{3}H-labelled deoxynucleotide triphosphate. Progress of the double-stranding can then be monitored in three ways: first, by the incorporation of the ^{3}H-label into trichloracetic acid-precipitable material; secondly, by the acquisition of resistance of the first strand to S1 nuclease digestion[34,35] and thirdly, by the decrease in electrophoretic mobility in a denaturing gel (Figure 9.2). After completion of the second reaction the plateau in the reaction is reached at 80% resistance of the ^{32}P counts to digestion by the nuclease.[36] Before the hairpin has been digested, the covalently linked first and second strands should have a combined molecular weight which is twice that of the first strand, and should migrate correspondingly slower in a fully denaturing gel. After S1 nuclease digestion the strands migrate independently in a denaturing gel and should have a mobility approximately the same as the single-stranded product of the first reaction. If a heterogeneous population of mRNAs is used as the template, these changes will be seen as shifts in the overall distribution pattern of the reaction products (Figure 9.2, lanes 6–8).

5. Enrichment for Sequences of Interest

The screening of cDNA clones for specific sequences can be a laborious procedure, as will be seen in Chapter 12. If the abundance of the sequence of interest in the ds-cDNA is only 0.01%, it will, on average, be found in only 1 in 10,000 of the clones generated. However, the number of clones which have to be screened can be markedly reduced by the inclusion of one or more steps which enrich the ds-cDNA for the desired sequences prior to cloning. There are a number of ways in which this can be achieved.

Enrichment of the mRNA

Appropriate selection of the biological material from which the mRNA is isolated is generally the first step towards obtaining an enriched mRNA population. The cells or tissue from which the mRNA is to be extracted should be chosen carefully to ensure that the mRNA of interest is as abundant as possible. When a gene is under hormonal or developmental control it is usual to find that the mRNA transcript is at its most abundant just before the peak in the rate of protein synthesis, but some initial studies may be required to determine the optimum sampling time. Sometimes mutants or natural variants may be available which possess increased levels of the mRNA: Lewis *et al.*, for example, cloned cDNA for dihydrofolate reductase from a line of Chinese hamster cells which overproduced the enzyme.[37]

A number of proteins, such as those which are secreted from the cell, are synthesised by polysomes which are bound to the endoplasmic reticulum. In this case considerable enrichment of the mRNA can be achieved by making a membrane preparation and extracting the mRNA from the membrane-bound polysomes.[38]

A commonly used method for mRNA enrichment is the fractionation of the mRNA by size. This is achieved either by sucrose gradient centrifugation[39] or by agarose gel electrophoresis.[40] The latter method gives better resolution of the RNA, but has the disadvantages that it is more laborious to carry out and generally gives lower yields of RNA. When considering whether or not size fractionation might be worthwhile it is worth remembering that it will be most effective for enriching for those mRNAs which are unusually small or large: that is, those mRNAs which code for polypeptides smaller than, say, 20×10^3 daltons, or larger than, say, 60×10^3 daltons. The fractions are monitored by translation of the RNA in one of the *in vitro* protein synthesising systems (Chapter 7). The *in vitro* translation products can be identified by their electrophoretic mobility and, if possible, by immunoprecipitation. At least a tenfold enrichment for the mRNA sequence of interest is to be expected after size-fractionation.

An ingenious method for selectively enriching an mRNA population for inducible sequences has recently been developed by Vitek *et al.*[41] Oligo(dT) cellulose (oligo(dT) which is covalently linked to cellulose powder) is used to prime the reverse transcription of mRNA from uninduced cells. After removing the template RNA, the mRNA population from induced cells is passed over a chromatography column made from the cellulose, which now has cDNA covalently linked to it. Sequences common to both types of cells will hybridise to the cDNA and those specific to the induced cells will pass through the column. The mRNA which still fails to bind after several passages over the column will therefore be greatly enriched for the inducible sequences. In principle, the method (called 'hybridisation-subtraction' chromatography) should be applicable to any situation where two very similar mRNA populations can be prepared which differ mainly by the presence of absence of the sequences(s) of interest. In practice the limitation of the technique is its requirement for large amounts (ca. 600 μg) of the 'uninduced' or 'depleted' mRNA population.

Enrichment of the cDNA

Occasionally procedures to enrich the cDNA or ds-cDNA itself have been employed. One such technique requires the digestion of the cDNA or ds-cDNA with restriction endonucleases. When the digestion products are analysed by electrophoresis and autoradiography it is possible to detect discrete restriction fragments generated from any cDNA species which represents at least 2% of the total mass of cDNA.[42] A band can be identified as being derived from a specific mRNA in several ways.[43] The most straightforward method is to reverse transcribe mRNA from a series of sucrose gradient fractions and then compare the intensity of the band(s) with the translational activity of the mRNA.

The cDNA may also be enriched for transcripts of a specific mRNA by fractionating it on alkaline sucrose gradients.[44] To be useful, however, this technique requires that the size of the mRNA is known and that a fairly high proportion of the transcripts are full-length. Unlike fractionation of mRNA itself there is usually no convenient way of assaying the cDNA fractions for the sequence of interest.

Using a technique analogous to hybridisation-subtraction chromatography of mRNA, Zimmermann *et al.* enriched their *Aspergillus* cDNA for differentiation-specific sequences by hybridising it to RNA from somatic cells.[45] The hybrids were removed by hydroxyapatite chromatography (which separates single-stranded DNA from double-stranded DNA), leaving unhybridised cDNA which had been selectively depleted of those sequences which occur in somatic cells. Like hybridisation-subtraction chromatography, however, this technique requires a considerable excess of the 'depleted' RNA.

6. Making Full-length Double-stranded cDNA

Sometimes priority may be given to obtaining cDNA clones which are full-length representations of the mRNA, despite the additional effort which this will entail. The chances of obtaining a full-length clone can be considerably improved by eliminating ribonuclease activity in the first reaction, as discussed above. However, further enrichment for complete copies of the mRNA can be achieved by fractionation of the ds-cDNA (after S1 nuclease digestion) on non-denaturing agarose or acrylamide gels. If the full-length ds-cDNA of interest is sufficiently abundant to be detected as a distinct band on an autoradiograph of the gel, or if its molecular weight is known, it can be excised and eluted from the gel.[46]

An alternative approach is first to construct and identify a partial cDNA clone representing at least the 3' terminal region of the mRNA. When excised and purified from the plasmid vector this cDNA fragment is then used to prime specifically the reverse transcription of the mRNA species of interest.[47] The extended cDNA fragment is then made double-stranded and re-cloned.

Even the most complete cDNA clones are invariably found to have lost 15–20 nucleotides corresponding to the extreme 5'-terminal region of the mRNA. These are the nucleotides removed when the hairpin loop is digested with S1 nuclease. To circumvent this problem, Land *et al.* have used a procedure which involves the elongation of the 3'-end of the cDNA with 20–25 dCMP residues.[48] Second-strand synthesis is then primed, not by the hairpin loop, but by a short oligo(dG) molecule which is hybridised to the oligo(dC) tail. Because the second strand is not covalently linked to the first strand there is then no need for the S1 nuclease digestion.

7. Applications of cDNA Cloning

To Identify Cloned Genomic DNA Fragments

As discussed in the introduction to this chapter, cDNA cloning is a critical intermediate step in the isolation of structural genes and their flanking sequences. The cloned cDNA fragment is radioactively labelled and used as a highly specific hybridisation probe to screen bacteriophage 'libraries' of nuclear DNA restriction fragments.[49]

To Determine the Primary Structure of a Protein

The techniques which have been developed in recent years by Maxam and Gilbert and by Sanger for the rapid sequencing of DNA (see Chapter 16) have made it theoretically possible for one person to obtain up to 1000 nucleotides of sequence in one week. Thus nucleotide sequencing of cloned cDNA fragments is a rapid and convenient means of determining the primary structure of a protein since the amino acid sequence can readily be deduced from the sequence of nucleotides. In cases where the protein is only available in very small amounts, or where there is difficulty in purifying it, this may be the only way that its amino acid sequence can be determined.

To Monitor the Abundance of mRNAs

It is often of considerable interest to estimate the abundance of a specific mRNA species, for example in the study of an inducible or a developmentally regulated gene, or in the comparison of normal and mutant strains. Cloned cDNA fragments provide an unrivalled means of assaying the abundance of an individual mRNA species, as demonstrated for example by Sim *et al.* in their study of chorion gene expression.[50]

To Analyse the Organisation of Nuclear Genes

Cloned cDNA sequences are often used to probe the nuclear genome to determine the number of gene copies and map the structural genes. Restriction digests of nuclear DNA, fractionated on agarose gels, are transferred to nitrocellulose filters (the 'Southern blot') and hybridised to a ^{32}P-labelled cDNA probe (Chapter 15).[51] The restriction fragments to which the cDNA hybridises are detected by autoradiography. If the nuclear DNA is digested with a series of different restriction endonucleases a restriction map of the structural genes and their flanking sequences can be deduced.

To Obtain Expression of a Eukaryotic Gene in Bacteria

Full-length cDNA sequences, when placed under the control of a bacterial promotor in one of the specially constructed 'expression' vectors,[52] can be used to programme the synthesis in bacteria of large amounts of a eukaryotic protein which is of particular scientific, medical or commercial interest. Two notable examples have been the successful bacterial syntheses of a human interferon[53] and the major antigen of foot and mouth disease.[54] The absence of introns in the cDNA sequence makes it a more suitable source of the genetic material than nuclear DNA for this purpose, since bacteria lack the RNA processing enzymes necessary to excise intervening sequences.

8. Conclusions

Although almost all methods of cDNA cloning are based on the procedure depicted in Figure 9.1, there is clearly considerable room for variation in the approach. The precise course of action will largely depend on factors such as the abundance of the mRNA of interest, its size and the desire for full-length cDNA clones. It is these factors which will determine how much time need be spent on the time-consuming procedures of enriching for the sequences of interest, optimising the enzyme reactions and purifying full-length transcripts. It can quite easily take several months to size fractionate the mRNA and optimise the enzyme reactions. On the other hand, once the RNA has been prepared and the reaction conditions have been established, the synthesis of the ds-cDNA can be completed within a week.

Further Reading

D.M. Glover, *Genetic Engineering: Cloning DNA* (Chapman and Hall, London, 1980).

R.W. Old and S.B. Primrose, *Principles of Gene Manipulation: An Introduction to Genetic Engineering*, 2nd ed (Blackwell Scientific Publications, Oxford, University of California Press, Berkeley and Los Angeles, 1981).

R.F. Schleif and P.C. Wensink, *Practical Methods in Molecular Biology*, (Springer-Verlag, New York).

J.K. Setlow and A. Hollaender (eds.), *Genetic Engineering: Principles and Methods*, (Plenum Press, New York), vols. 1 and 2.

R. Williamson (ed.), *Genetic Engineering*, (Academic Press, New York) vols. 1 and 2.

References

1. T. Maniatis, A. Efstratiadis, S.G. Kee and F.C. Kafatos, 'In vitro synthesis and molecular cloning of eukaryotic structural genes', in D.P. Nierlich, W.J. Rutter and C.F. Fox (eds.), *Molecular Mechanisms in the Control of Gene Expression*, E.C.M.-U.C.L.A. Symposia on Molecular and Cullular Biology Vol. V. (Academic Press, New York, 1976), 513–33.

2. A. Efstratiadis and L. Villa-Komaroff, 'Cloning of double-stranded cDNA', in J.K. Setlow and A. Hollaender (eds.), *Genetic Engineering: Principles and Methods*, (Plenum Press, New York, 1979), 15–36.
3. Maniatis, Efstratiadis, Kee and Kafatos, 'In vitro synthesis'.
4. H.M. Goodman and R.J. MacDonald, 'Cloning of hormone genes from a mixture of cDNA molecules', *Methods in Enzymology, 68* (1979), 75–90.
5. D. Baltimore, 'RNA-dependent DNA polymerase in virions of RNA tumour viruses', *Nature, 226* (1970), 1209–11.
6. H. Temin and S. Mizutani, 'RNA-dependent DNA polymerase in virions of Rous sarcoma virus', *Nature, 226* (1970), 1211–13.
7. Efstratiadis and Villa-Komaroff, 'Cloning of double-stranded cDNA'.
8. Ibid.
9. P. Humphries, M. Cochet, A. Krust, P. Gerlinger, P. Kourilsky and P. Chambon, 'Molecular cloning of extensive sequences of the in vitro synthesised chicken ovalbumin structural gene', *Nucleic Acids Res., 4* (1977), 2389–406.
10. G.N. Buell, M.P. Wickens, F. Payvar and R.T. Schimke, 'Synthesis of full length cDNAs from four partially purified oviduct mRNAs', *J. Biol. Chem., 253* (1978), 2471–82.
11. F. Payvar and R.T. Schimke, 'Methylmercury hydroxide enhancement of translation and transcription of ovalbumin and conalbumin mRNAs', *J. Biol. Chem., 254* (1979), 7636–42.
12. Efstratiadis and Villa-Komaroff, 'Cloning of double-stranded cDNA'.
13. Buell, Wickens, Payvar and Schimke, 'Synthesis of full length cDNAs'.
14. Efstratiadis and Villa-Komaroff, 'Cloning of double-stranded cDNA'.
15. Maniatis, Efstratiadis, Kee and Kafatos, 'In vitro synthesis'.
16. M.P. Wickens, G.N. Buell and R.T. Schimke, 'Synthesis of double-stranded DNA complementary to lysozyme, ovomucoid and ovalbumin mRNAs: optimization for full length second strand synthesis by *Escherichia coli* DNA polymerase I', *J. Biol. Chem., 253* (1978), 2483–95.
19. T.E. Shenk, C. Rhodes, P.W.J. Rigby and P. Berg, 'Biochemical method for mapping mutational alterations in DNA with S1 nuclease: the location of deletions and temperature sensitive mutations in Simian virus 40', *Proc. Natl. Acad. Sci. USA., 72* (1975), 989–93.
20. L.A. McReynolds, J.F. Catterall and B.W. O'Malley, 'The ovalbumin gene: cloning of a complete ds–cDNA in a bacterial plasmid', *Gene, 2* (1977), 217–31.
21. J. Leis and J. Hurwitz, 'RNA-dependent DNA polymerase from avian myeloblastosis virus', *Methods in Enzymology, 29* (1974), 143–50.
22. D.L. Kacian and J.C. Myers, 'Synthesis of extensive, possibly complete, DNA copies of poliovirus RNA at high yields and at high specific activities', *Proc. Natl. Acad. Sci. USA., 73* (1976), 2129–5.
23. G. de Martynoff, E. Pays and G. Vassart, 'Synthesis of a full-length DNA complementary to thyroglobulin 33S messenger RNA', *Biochem. Biophys. Res. Commun., 93* (1980), 645–53.
24. J.L. Berge-Lefranc, G. Cartouzou, Y. Maltiery, F. Perrin, B. Jarry and S. Lissitzky, 'Cloning of four DNA fragments complementary to human thyroglobulin messenger RNA', *Eur. J. Biochem., 120* (1981), 1–7.
25. Buell, Wickens, Payvar and Schimke, 'Synthesis of full length cDNAs'.
26. Maniatis, Efstratiadis, Kee and Kafatos, 'In vitro synthesis'.
27. Goodman and MacDonald, 'Cloning of hormone genes'.
28. Wickens, Buell and Schimke, 'Synthesis of double-stranded DNA'.
29. Goodman and MacDonald, 'Cloning of hormone genes'.
30. M.W. McDonnell, M.N. Simon and W.F. Studier, 'Analysis of restriction fragments of T7 DNA and determination of molecular weights by electrophoresis in neutral and alkaline gels', *J. Mol. Biol., 110* (1977), 119–46.
31. T. Maniatis and A. Efstratiadis, 'Fractionation of low molecular weight DNA or RNA in polyacrylamide gels containing 98% formamide', *Methods in Enzymology, 65* (1980) 299–305.
32. Ibid.
33. B.G. Forde, M. Kreis, M.B. Bahramian, J.A. Matthews, B.J. Miflin, R.D. Thompson, D. Bartels and R.B. Flavell, 'Molecular cloning and analysis of cDNA sequences derived from poly A⁺ RNA from barley endosperm: identification of B Hordein related clones',

Nucleic Acids Res., *9* (1981), 6689–6707.

34. Goodman and MacDonald, 'Cloning of hormone genes'.
35. Wickens, Buell and Schimke, 'Synthesis of double-stranded DNA'.
36. Ibid.
37. J.A. Lewis, D.T. Kurtz and P.W. Melera, 'Molecular cloning of Chinese hamster dihydrofolate reductase-specific cDNA and the identification of multiple dihydrofolate reductase mRNAs in antifolate-resistance Chinese hamster lung fibroblasts', *Nucleic Acids Res.*, *9* (1981), 1331–22.
38. D. Baulcombe and D.P.S. Verma, 'Preparation of a complementary DNA for leghaemoglobin and direct demonstration that leghaemoglobin is encoded by the soybean genome', *Nucleic Acids Res.*, *5* (1978), 4141–53.
39. B. Mach, C.H. Faust and P. Vassalli, 'Purification of 14S messenger RNA of immunoglobulin light chain that codes for a possible light-chain precursor', *Proc. Natl. Acad. Sci. USA.*, *70* (1973), 451–5.
40. J.R. Bedbrook, S.M. Smith and R.J. Ellis, 'Molecular cloning and sequencing of cDNA encoding the precursor to the small subunit of the chloroplast enzyme ribulose-1, 5-bisphosphate carboxylase', *Nature*, *287* (1980), 692–7.
41. M.P. Vitek, S.G. Kreissman and R.H. Gross, 'The isolation of ecdysterone inducible genes by hybridization subtraction chromatography', *Nucleic Acids Res.*, *9* (1981), 1191–202.
42. Goodman and MacDonald, 'Cloning of hormone genes'.
43. Ibid.
44. H. Diggelman, C.H. Faust and B. Mach, 'Enzymatic synthesis of DNA complementary to purified 14S messenger RNA of immunoglobulin light chain', Proc. Natl. Acad. Sci. USA., *70* (1973), 693–6.
45. C.R. Zimmermann, W.C. Orr, R.F. Leclerc, E.C. Bernard and W.E. Timberlake, 'Molecular cloning and selection of genes regulated in Aspergillus development', *Cell*, *21* (1980), 709–15.
46. Maniatis, Efstratiadis, Kee and Kafatos, 'In vitro synthesis'.
47. M.J. Gethin, J. Bye, J. Skehel and M. Waterfield, 'Cloning and DNA sequence of double-stranded copies of haemaglutinin genes from H2 and H3 strains elucidates antigenic shift and drift in human influenza virus', *Nature*, *287* (1980), 301–6.
48. H. Land, M. Grez, H. Hauser, W. Lindenmaier and G. Schutz, '5'-Terminal sequences of eukaryotic mRNA can be cloned with high efficiency', *Nucleic Acids Res.*, *9* (1981), 2251–66.
49. T. Maniatis, R.C. Hardison, E. Lacy, J. Lauer, C. O'Connell, D. Quon, G.K. Sim and A. Efstratiadis, 'The isolation of structural genes from libraries of eukaryotic DNA', *Cell*, *15* (1978), 687–701.
50. G.K. Sim, F.C. Kafatos, C.W. Jones, M.D. Koehler, A. Efstratiadis and T. Maniatis, 'Use of a cDNA library for studies on evolution and developmental expression of the chorion multigene family', *Cell*, *18* (1979), 1303–16.
51. A.J. Jeffreys and R.A. Flavell, 'A physical map of the DNA regions flanking the rabbit β-globin gene', *Cell*, *12* (1977), 429–39.
52. J.F. Morrow, 'Recombinant DNA techniques', *Methods in Enzymology*, *68* (1979), 3–24.
53. S. Nagata, H. Taira, A. Hall, L. Johnsrud, M. Streuli, J. Escodi, W. Boll, K. Cantell and C. Weissmann, 'Synthesis in *E. coli* of a polypeptide with human leukocyte interferon activity', *Nature*, *284* (1980), 316–20.
54. H. Kupper, W. Keller, C. Kurz, S. Forss, H. Schaller, R. Franze, K. Strohmaier, O. Marquardt, V.G. Zaslavsky and P.H. Hofschneider, 'Cloning of cDNA of major antigen of foot and mount disease virus and expression in *E. coli*', *Nature*, *289* (1981), 555–9.

10 PLASMID ISOLATION

Stephen A. Boffey

CONTENTS

1. Introduction

Some important bacterial genes are not located in the main chromosomal DNA but in independently replicating molecules of circular, double-stranded, 'plasmid' DNA. Such genes include those for antibiotic resistance, antibiotic synthesis, toxin production, nitrogen fixation, production of degradative enzymes, and conjugation; so plasmids are obviously of interest in their own right. However, in the context of this book, plasmids are mainly of interest as 'vectors' for the cloning of DNA molecules. As will be seen in chapters 11 and 12, it is possible to obtain large quantities of a particular DNA (e.g. cDNA – see Chapter 9) by inserting it into plasmid DNA, to give enlarged plasmids ('vector' plasmids) which can be introduced into a suitable host bacterium in which they will replicate. Culture of the cells will result in the production of more plasmid DNA, which can then be isolated from the cells and the inserted DNA recovered.

Most of the plasmids used as cloning vectors are not naturally occurring forms, but have been extensively modified so that they have properties useful for cloning. Many such plasmids have been produced, and they are usually identified by a code of the form pAB123, where 'p' stands for 'plasmid', 'AB' are initials identifying the worker or laboratory which isolated the plasmid, and the number is the laboratory's code for that particular plasmid.

The choice of plasmid and host will depend on the answers to several questions: for example – how large is the DNA fragment which is to be cloned; is it hoped to obtain 'expression' of inserted DNA (i.e. to produce an RNA or protein product coded for by the inserted DNA, using the host's transcription and translation systems); could any plasmid-containing hosts potentially be dangerous? Information in Bolivar and Backman[1] and Kahn et al.[2] should help the reader choose an appropriate plasmid.

For most DNA cloning work the plasmid should contain markers to aid its detection in transformed cells (e.g. antibiotic resistance genes), single cleavage sites for a number of restriction endonucleases to aid insertion of DNA fragments and, especially if the inserted DNA is not expressed, its insertion should inactivate a marker gene to allow selection of hosts containing hybrid plasmids. Naturally the plasmid must be replicated in the host cell, and if its replication is 'relaxed' (i.e. not stringently coupled to chromosomal DNA replication) there is the possibility of increasing the number of plasmid copies per cell by selectively inhibiting chromosomal DNA replication, thus improving yields of plasmid DNA and enriching it with respect to chromosomal DNA.

The isolation of pure, intact, plasmid DNA in high yields is made simpler if the plasmid is small. Small plasmids are relatively resistant to shearing and will therefore remain in their native, supercoiled, covalently closed circular (CCC) form, whilst the high molecular weight chromosomal DNA will be broken into large linear fragments. These differences in size and shape can be exploited, as described later, to separate the two types of DNA from each other.

Desirable characteristics of the host bacterium include ease of trans-formation, of culture, and of lysis. The combination of the plasmid pBR322 (molecular weight 2.7×10^6 daltons) and the host *Escherichia coli* meets all the above requirements, and this is consequently one of the most widely used systems for the cloning of DNAs.

Plasmids must be isolated from the host prior to insertion of DNA fragments and subsequent cloning, after which the vector plasmid with its inserted sequences must be isolated in order to recover the cloned DNA fragments. In this chapter I shall describe some of the most popular methods used for the isolation of small plasmids (molecular weight $< 10 \times 10^6$) from bacterial hosts, although many of the techniques can also be applied to the isolation of other classes of extrachromosomal DNA from other organisms.

2. Methods

Amplification of Plasmid

Isolation of pure plasmid in high yields will obviously be easier if the plasmid forms a large proportion of the cellular DNA prior to isolation. This can occur when the plasmid replication is 'relaxed', its rate of initiation not being coupled to that of chromosomal DNA, thus giving many plasmid copies per cell. Much greater amplification of the plasmid can be obtained by treating cells in late log phase with an inhibitor of protein synthesis, such as chloramphenicol or, if the plasmid carries a chloramphenicol resistance marker, spectinomycin. Because initiation of chromosomal DNA replication depends on continued protein synthesis, this replication will stop; however, it has been found that a 'relaxed' plasmid such as pBR322 can continue to replicate in the absence of protein synthesis, thus producing up to 3000 copies per cell after an overnight incubation in the presence of antibiotic.[3]

Extraction of Plasmid DNA

Although there are scores of recipes for extracting plasmid DNA from bacteria, most of them make use of the difference in sizes of plasmid and chromosomal DNAs to give preferential release of plasmid. In the case of gram negative bacteria such as *E. coli* it is common to treat the cells with a Tris/EDTA buffer to remove the outer envelope and so expose the peptidoglycan of the cell wall, which is then partially digested in isoosmotic solution using lysozyme. Finally the plasma membrane is gently disrupted using a detergent such as sodium dodecyl sulphate (SDS). In cases where the plasmid is to be used for transformation of an SDS-sensitive bacterium (e.g. *E. coli* χ 1776) sodium deoxycholate and polyoxyethylene-20-cetyl-ether (Brij-58) can be used in place of SDS.[4,5] This lysis of the bacteria is visible as a clearing of the originally cloudy suspension, and should result in a semi-gelled solution, in which the gel is made up of a tangle of very high molecular weight chromosomal DNA and cell wall debris. By high-speed centrifugation the gel can be packed down to form a translucent pellet containing most of the chromosomal DNA, leaving plasmid DNA, RNA, carbohydrate, protein and some chromosomal DNA in the supernatant (the 'cleared lysate'). An alternative method which is particularly useful for very small scale work, using as little as one 3 mm diameter colony of cells,[6] is to lyse the cells with Tris/EDTA/lysozyme/SDS as above, but with the addition of NaOH to the SDS, giving a pH between 12.0 and 12.5. At this pH DNA is denatured, resulting in the separation of strands of fragmented chromosomal DNA; however, the two strands of plasmid DNA remain entangled and so will renature when the solution is neutralised. This is done by adding an excess of concentrated acidic sodium acetate solution, with the result that the chromosomal DNA, high molecular weight RNA, and proteins (as complexes with SDS) precipitate out, and can be removed by centrifugation to give only plasmid DNA and a lot of low molecular weight RNA in the supernatant. The selective denaturation of chromosomal DNA can also be achieved by very brief boiling, but the timing of this treatment is critical.[7]

Purification of Plasmid

The degree of purity needed will depend on the experiments to be performed on the plasmid. Chromosomal DNA may interfere with transformation, hybridisation and gel electrophoresis; high molecular weight RNA will inhibit restriction endonucleases; proteins may include nucleases, which could degrade the plasmid; salts and detergents carried over from the isolation medium may be unsuitable for subsequent steps. So usually some cleaning-up of the plasmid will be needed after its isolation from the bacterium.

Deproteinisation and Removal of RNA. Proteins can be removed by extraction with a phenol/chloroform mixture, which both denatures proteins and concentrates them at the aqueous/organic interface, thus deproteinising the nucleic acid preparation. Repeated extraction may be needed to remove all protein, and the easiest way to monitor deproteinisation is by measuring the ratio of absorbances at 260 nm and 280 nm: pure nucleic acids should have an A_{260}/A_{280} ratio of 2.0.[8]

There are several ways of dealing with RNA, depending on the isolation method employed, and the intended use of the plasmid. The RNA recovered after alkaline denaturation is of low molecular weight, and does not inhibit restriction endonucleases, so it is usually unnecessary to remove it. Such preparations can be treated with two volumes of chilled ethanol to precipitate the total nucleic acids. After two or three ethanol precipitations the sample can be redissolved in an appropriate buffer, and is suitable for gel electrophoresis, restriction or transformation.[9] Other methods of plasmid preparation usually result in contamination with higher molecular weight RNA, which inhibits restriction enzymes. This can be destroyed by digestion with RNase (heat-treated to inactivate any DNase contaminants), or can be separated from the plasmid DNA by gel filtration.[10]

The following techniques for separating plasmids from chromosomal DNA also remove most RNA, and so can be used in place of RNase treatment or gel filtration.

Isopycnic Ultracentrifugation. The base compositions, and hence densities, of plasmid and chromosomal DNAs are normally so similar that they cannot be separated by virtue of a density difference. However, it is possible to produce such a difference using the fluorescent dye ethidium bromide, which complexes with DNA by intercalation between base pairs, thus reducing its buoyant density. Intercalation of dye can only occur if the DNA double helix unwinds slightly, a process which results in strain in CCC DNA but which is unhindered in linear or open-circular molecules. Consequently, in saturating concentrations of ethidium bromide CCC plasmid DNA will bind less dye per unit length, and will have a higher buoyant density, than linear chromosomal DNA of the same base composition.

It is therefore possible to separate CCC plasmid DNA from other DNA molecules by isopycnic ultracentrifugation in a CsCl density gradient in the presence of excess ethidium bromide. The reader is referred to Schleif and Wensink[11] for practical details of CsCl density gradient ultracentrifugation. After 40 hours at 38,000 rpm (125,000 g av), 15°C, in a conventional swing-out or angle rotor, CCC plasmid DNA will have formed a band below the linear DNA molecules (Figure 10.1a). DNA bands are visible when viewed by u.v. light owing to the red fluorescence of the ethidium bromide bound to them. RNA, which also binds some dye, will have moved to the bottom of the centrifuge tube, and non-fluorescent polysaccharide bands are also often seen near to the DNA. Plasmid DNA can be recovered from the centrifuge tube by piercing the side of the tube just beneath the plasmid band, by piercing the base of the tube and collecting fractions as they drip out, by displacing the gradient using a dense solution, or by carefully drawing off the plasmid band using a syringe. It should be noted that ethidium bromide may be carcinogenic, so must be handled with care.

This method, especially if repeated, can produce very pure CCC plasmid DNA, free of chromosomal DNA, RNA and carbohydrate. The ethidium bromide can be removed by repeated extraction of the DNA solution with iso-amyl alcohol, or iso-propyl alcohol saturated with CsCl, and CsCl is eliminated by dialysis or gel filtration. Propidium iodide is sometimes used instead of ethidium bromide, and is claimed to give increased band separation, but it is very expensive.[12] Probably the main disadvantage of CsCl density gradient purification of DNA is that it is so slow, needing at least 40 hours of ultracentrifugation. In fact the CsCl gradient is established within a couple of hours, but the average distance which must be covered by a DNA molecule as it moves through CsCl solution to its equilibrium position is quite large, hence the need for long centrifugations to achieve

Figure 10.1: Density Gradient Ultracentrifugation of Nucleic Acids

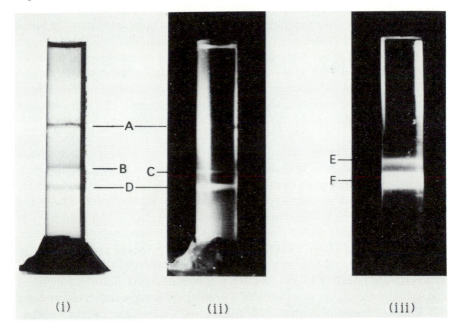

(i)   (ii)   (iii)

Note: a. A cleared lysate was obtained from 1l of *E. coli* ω310 cells containing the plasmid YRp12, amplified by treatment with chloramphenicol. After deproteinisation using phenol/chloroform (1:1, v/v), nucleic acids were precipitated using ethanol. The precipitate was redissolved in 3.6 ml of 15 mM NaCl, 1.5 mM sodium citrate to which were added 3.8 g CsCl and 0.4 ml ethidium bromide (5 mg/ml); this solution was then transferred to 3 centrifuge tubes. After centrifugation at 38,000 rpm (125,000 g_{av}) for 40h at 15°C the tubes were viewed by visible (i) and u.v. (ii) light to reveal several bands of material. 'A' non-fluorescent; 'B' non-fluorescent carbohydrate; 'C' fluorescent chromosomal, OC and linear plasmid DNAs; 'D' fluorescent CCC plasmid DNA. Note the success of the cleared lysate method in producing a preparation highly enriched in plasmid DNA; there is so much plasmid that it concentrates ethidium bromide in band 'D', making it visible even without u.v. illumination.

b. The dye Hoechst 33258 was used to separate two classes of DNA molecules which differed slightly in their AT contents, in this case yeast mitochondrial and nuclear DNAs. Centrifugation was as described in (a), using Hoechst 33258 in place of ethidium bromide. When viewed by u.v. light two bands were seen (iii): 'E' – the less dense mitochondrial DNA, and 'F' nuclear DNA.

(Samples kindly prepared by Paul Chambers, The Hatfield Polytechnic.)

equilibrium. This problem can be overcome by the use of a step gradient (concentrated CsCl beneath aqueous DNA) in a reorientating vertical rotor.[13] The step gradient results in rapid movement of all DNA to the centre of the tube, and the rotor geometry ensures that no molecule has far to travel to its equilibrium position, with the bonus of high capacity and excellent band separation. The net result is that good resolution of DNA bands can be obtained after only two hours using a vertical rotor.

In those cases where there is even a small difference of buoyant density between plasmid and chromosomal DNA, the difference can be exaggerated using the dye Hoechst 33258. This dye reduces the density of DNA to which it is bound, and, because it binds preferentially to AT-rich regions, will tend to reduce the density of DNA in proportion to its AT content, thus enhancing any natural differences in buoyant density between two DNA species. Consequently CsCl density gradient ultracentrifugation with Hoechst 33258 can separate DNA molecules on the basis of their AT contents (Figure 10.1b). This method is particularly useful when the plasmid is too large to resist shearing during even the gentlest isolation procedure, and so cannot be isolated intact, ruling out the use of CsCl-ethidium bromide centrifugation.[14]

Chromatography. Even when time can be saved by using a vertical rotor, ethidium bromide and CsCl must be removed from the DNA, and this adds to the inconvenience of density gradient techniques. However, it is possible to replace CsCl ultracentrifugation with adsorption chromatography, using hydroxyapatite or RPC-5 as the chromatography matrix.

Hydroxyapatite chromatography. Hydroxyapatite is crystalline calcium phosphate, $(Ca_2(PO_4)_2)_3 Ca(OH)_2$, and will bind proteins and nucleic acids at low phosphate concentrations. If the phosphate concentration is progressively raised, proteins and RNA are eluted first, followed by small DNA molecules (e.g. plasmid DNA) and finally high molecular weight DNA at the highest phosphate concentrations. Therefore, if a crude nucleic acid preparation (e.g. a cleared lysate) is loaded onto a column of hydroxyapatite at moderate phosphate concentrations, only DNA will be adsorbed, all protein, RNA and detergent passing straight through with the eluting buffer. Elution using a step or continuous concentration gradient of phosphate will allow plasmid and chromosomal DNAs to be collected separately.[15] This one procedure removes contaminants from plasmid DNA, can concentrate the plasmid (the *volume* loaded does not matter, only the total amount of nucleic acids), and it has a high capacity (e.g. it is claimed that l g dry weight of hydroxyapatite will bind 1.5 mg DNA[16]). The drawback of the technique is that maximum flow rates tend to be very low, owing to the tight packing of the crystals, and great care in handling is needed to produce reproducible results; it is not advisable to re-use such columns, as their flow properties tend to deteriorate during use.

To prevent this tight packing of the hydroxyapatite crystals, they can be entrapped in beads of agarose gel (LKB supply such a preparation as 'HA-Ultrogel'). Relatively high flow rates can be used through columns prepared from these beads, and the pore size of the agarose gel is large enough to allow easy access of all except the largest molecules to the hydroxyapatite. For most protein fractionations hydroxyapatite-agarose can be used in the same way as hydroxyapatite slurries. However, the gel exclusion limit, although large, will be low enough to exclude high molecular weight double-stranded DNA, such as the chromosomal DNA fragments present in a cleared lysate, with the result that such DNA can only bind to those few crystals present at the surface of the agarose beads. Hence the column has a low adsorption capacity for high molecular weight DNA, which can be eluted with low phosphate concentrations.

Supercoiled CCC plasmid DNA, being very compact, will not be excluded by the agarose, and so will bind strongly to the column. Using hydroxyapatite-agarose columns, protein, RNA and chromosomal DNA can be eluted at low ionic concentrations, allowing pure plasmid DNA to be recovered using a higher ionic strength. Such columns can be re-used many times.[17] For a few applications where high phosphate concentrations are undesirable (e.g. use of alkaline phosphatase or polynucleotide kinase) it may be necessary to dialyse the purified plasmid, or an alternative purification process may be used.

RPC-5 Chromatography. In 1971, Pearson *et al.*[18] described the separation of tRNAs using a reversed-phase chromatography column they called RPC-5, which was supported on polychlorotrifluoroethylene, and used a trialkylmethylammonium exchanger. This column was also found to be well suited to the separation of DNA fragments by HPLC, being a preparative alternative to agarose gel electrophoresis.[19]

For DNA molecules up to a few million daltons molecular weight separation is mainly on the basis of size, although base composition and 'sticky' ends (short, complementary lengths of single-stranded DNA at each end of the molecule, often produced by restriction endonucleases; see Chapter 11) will also affect binding to the column. In general, nucleic acids of increasing size are sequentially eluted using a gradient of increasing salt concentration, and so plasmid DNA is eluted after RNA but before chromosomal DNA. Owing to the limited availability of RPC-5, Bethesda Research Laboratories have developed another chromatography matrix with almost identical properties, called RPC-5 ANALOG; this material can be used to separate plasmid from up to 1 mg nucleic acids/1.7 g matrix in an HPLC column. Plasmid DNA can be purified in a 'batch' process using RPC-5 ANALOG; gel filtration is used to remove low molecular weight nucleic acids, and then the higher molecular weight nucleic acid mixture is adsorbed to RPC-5 ANALOG matrix by mixing in a buffer of low ionic strength. Stepwise elution of RNA, plasmid DNA, and chromosomal DNA is achieved by washing the matrix with buffers of increasing salt concentration, the exact concentrations being chosen to suit the particular plasmid and host DNA.[20] Although this method is more labour-intensive than HPLC, owing to the need for repeated centrifugation at each elution step, it is quicker, and does not require expensive chromatographic equipment.

Preparative Electrophoresis. Plasmid molecules, especially in their compact, superhelical, covalently closed circle form, can easily be separated from RNA and large fragments of chromosomal DNA by electrophoresis in agarose gels. Using preparative gel electrophoresis equipment, milligram quantities of plasmid can be isolated from a mixture of nucleic acids; however, it may not be possible to achieve complete separation between plasmid and small fragments of chromosomal DNA. Chapter 14, Section 5, contains further information about preparative electrophoresis.

Analysis of Plasmid Preparations

The purity of a plasmid preparation is most easily checked by agarose gel electrophoresis (Chapter 14), which will reveal the presence of RNA and chromosomal DNA, and will also show if the plasmid is in its CCC, open circle or linear forms (Figure 14.1, Chapter 14).

Gel electrophoresis can also be used to determine the molecular weight of a plasmid (Chapter 14), as long as it has been converted to its linear form by treatment with a restriction endonuclease (Chapter 11) which cuts at only one site on the plasmid. Restriction endonucleases which cleave at a few sites will produce several DNA fragments of various lengths, and these will form a characteristic pattern of bands after electrophoresis. This type of analysis will also reveal if DNA has been inserted into the plasmid, as such inserted sequences will change the molecular weight of the fragment they are inserted in, producing one or more new bands on electrophoresis, depending on whether or not the inserted sequence is itself cleaved (see Chapter 11 for more details).

References

1. F. Bolivar and K. Backman, 'Plasmids of *Escherichia coli* as cloning vectors', *Methods in Enzymology*, *68* (1979), 245–67.
2. M. Kahn, R. Kolter, C. Thomas, D. Figurski, R. Meyer, E. Remaut and D.R. Helinski, 'Plasmid cloning vehicles derived from plasmids ColE1, F, R6K, and RK2', *Methods in Enzymology*, *68* (1979), 268–80.
3. R.W. Old and S.B. Primrose, *Principles of gene manipulation*. 2nd edn. Studies in microbiology, *2* (Blackwell, Oxford/University of California Press, Berkeley and Los Angeles, 1981), 33.
4. D.B. Clewell and D.R. Helinski, 'Properties of a supercoiled deoxyribonucleic acid-protein relaxation complex and strand specificty of the relaxation event', *Biochemistry*, *9* (1970), 4428–40.
5. M.V. Norgard, 'Rapid and simple removal of contaminating RNA from plasmid DNA without the use of RNase', *Anal. Biochem.*, *113* (1981), 34–42.
6. H.C. Birnboim and J. Doly, 'A rapid alkaline extraction procedure for screening

recombinant plasmid DNA', *Nucleic Acids Res.*, 7 (1979), 1513–23.

7. D.S. Holmes and M. Quigley, A rapid boiling method for the preparation of bacterial plasmids. *Anal. Biochem.*, *114* (1981), 193–7.
8. E. Layne, 'Spectrophotometic and turbidimetric methods for measuring proteins', *Methods in Enzymology*, *3* (1957), 447–54.
9. Birnboim and Doly, 'Rapid alkaline extraction'.
10. Norgard, 'Rapid and simple removal of contaminating RNA'.
11. R.F. Schleif and P.C. Wensink, *Practical methods in molecular Biology* (Springer-Verlag, New York, 1981).
12. Norgard, 'Rapid and simple removal of contaminating RNA'.
13. J.R. Wells and C.F. Brunk, 'Rapid CsCl gradients using a vertical rotor', *Anal. Biochem.*, *97* (1979), 196–201.
14. M. Fennewald, W. Prevatt, R. Meyer and J. Shapiro, 'Isolation of Inc P-2 plasmid DNA from *Pseudomonas aeruginosa*', *Plasmid*, *1* (1978), 164–73.
15. M. Shoyab and A. Sen, 'The isolation of extrachromosomal DNA by hydroxyapatite chromatography', *Methods in Enzymology*, *68* (1979), 199–206.
16. BDH, Poole, England, 'Hydroxyapatite for nucleic acid work', Applications pamphlet 1227MP/5.0/1079.
17. LKB, Bromma, Sweden, 'HA-Ultrogel hydroxyapatite-agarose gel for adsorption chromatography', Instruction Manual.
18. R.L. Pearson, J.F. Weiss and A.D. Kelmers, 'Improved separation of transfer RNAs on polychlorotrifluoroethylene-supported reversed-phase chromatography columns', *Biochim. Biophys. Acta*, *228* (1971), 770–4.
19. R.D. Wells, S.C. Hardies, G.T. Horn, B. Klein, J.E. Larson, S.K. Neuendorf, N. Panayotatos, R.K. Patient and E. Selsing, *Methods in Enzymology*, *65* (1980), 327–47.
20. Bethesda Research Laboratories Inc., Gaithersburg, USA, BRL Catalogue (1981), 86.

11 THE USE OF RESTRICTION ENDONUCLEASES AND T4 DNA LIGASE

Frits R. Mooi and Wim Gaastra

CONTENTS

1. The Use of Restriction Endonucleases

Introduction

Bacteria possess restriction-modification systems, each system of which consists of a modification methylase and a restriction endonuclease. The modification methylase recognises a specific sequence and methylates certain adenines or cytosines within this sequence. The corresponding restriction endonuclease recognises the same sequence and in most cases cleaves the DNA when the sequence has not been methylated by the modification methylase.

Restriction-modification systems probably prevent foreign DNA (having a foreign methylation pattern) from functioning within the cell by causing double-stranded scissions in the foreign DNA at a limited number of sites.

Currently, three types of restriction-modification systems are recognised, Types I, II and III.[1,2] Since only the Type II restriction endonucleases are used extensively in recombinant DNA technology, only they will be considered in this chapter (see Roberts[3] for a recent compilation). Type II restriction endonucleases recognise a specific sequence of four to six base pairs, which generally has a twofold symmetry (Table 11.1). Usually the restriction endonuclease causes a double-stranded scission within this sequence. The double-stranded scission may be blunt-ended (e.g. *Sma* I, Table 11.1) or staggered (e.g. *Eco*R I, Table 11.1). In the latter case the ('cohesive or sticky') ends are readily rejoined by base pairing with the complementary ends of other DNA fragments generated by cleavage with the same enzyme. Staggered scissions may result in 3'- or 5'-single-stranded extensions (e.g. *Pst* I and *Eco*R I respectively, Table 11.1).

Some restriction endonucleases (e.g. *Sfa*N I, Table 11.1) do not cleave the DNA within their recognition sequence, but some bases away from it. In these cases the scission is not introduced in a specific DNA sequence and if single-stranded extensions are generated, the ends cannot be rejoined to those of other DNA fragments generated by the same enzyme, because they are not complementary. Restriction endonucleases that cleave within their recognition sequence can also generate heterogeneous ends, because one base-pair within the sequence may vary (e.g. *Hin*f I, Table 11.1). Restriction endonucleases that create heterogeneous single-stranded extensions are generally not suitable for cloning experiments.

Most known restriction endonucleases are inactive when specific

199

Table 11.1: Recognition Sequences of and Ends Generated by some Restriction Endonucleases

Restriction endonuclease	Recognition sequence[a]	Ends created
*Sma*I	5'-CCC↓GGG-3' 3'-GGG↑CCC-5'	-CCC-3' + 5'-GGG- -GGG-5' 3'-CCC-
*Eco*R I	5'-G↓AATTC-3' 3'-CTTAA↑G-5'	-G-3' + 5'-AATTC- -CTTAA-5' 3'-G-
Pst I	5'-CTGCA↓G-3' 3'-G↑ACGTC-5'	-CTGCA-3' + 5'-G- -G-5' 3'-ACGTC-
*Sfa*N I	5'-N↓NNNNNNNNNNGATGC-3' 3'-NNNNN↑NNNNNCTACG-5'	-N-3' + 5'-NNNNNNNNNNGATGC- -NNNNN-5' 3'-NNNNNCTACG
Hinf I	5'-G↓ANTC-3' 3'-CTNA↑G-5'	-G-3' + 5'-ANTC- -CTNA-5' 3'-G-

Note: a. The mode of cleavage is indicated by the arrows.

methylated cytosine or adenine residues are present in their recognition sequence. Therefore caution is necessary when using some restriction endonucleases to study DNA that has an uncharacterised methylation pattern. *Escherichia coli* has three (chromosome-encoded) methylation systems, which function independently: the methylase which is part of the restriction-modification system, the adenine methylase (*dam* gene product) and the cytosine methylase (*dcm* gene product). The *dam* methylase recognises the sequence GATC in double-stranded DNA, and methylates both adenines in it.[4] The *dcm* methylase recognises the sequence CCXGG, and methylates the two interior located cytosines in it.[5,6] DNA modified by the *dam* methylase is not cleaved by *Mbo* I (Table 11.2) because this restriction endonuclease recognises the same sequence as the *dam* methylase but does not cleave it when the adenine is methylated. DNA modified by the *dcm* methylase is not cleaved by *Eco*R II for the same reason (Table 11.2).

It is also possible that only certain recognition sites of a restriction endonuclease are protected by *dam* or *dcm* methylation. For example *dam* methylation protects against *Taq* I (Table 11.2) only if it is part of the sequence

$$5'\text{-T C G}\textcircled{A}\text{T C-3'} \qquad \text{or} \qquad 5'\text{-G}\textcircled{A}\text{T C G A-3'}$$
$$3'\text{-A G C T}\textcircled{A}\text{G-5'} \qquad\qquad 3'\text{-C T}\textcircled{A}\text{G C T-5'}$$

where methylated bases are encircled. To circumvent these problems one can use restriction endonucleases that cleave these methylated sequences (for example *Sau*3A I can be used instead of *Mbo* II; and *Bst*N I instead of *Eco*R II, see Table 11.2). Alternatively one can use *dam⁻* or *dcm⁻* strains or strains which are deficient in restriction and modification (r⁻m⁻). It should be noted however, that in most cases methylation of DNA does not create serious problems and can even be used to selectively protect certain sites

Table 11.2: The Influence of Methylation on Cleavage by Restriction Endonucleases[6]

Restriction endonuclease	Recognition[a] sequence	Methylated[b] sequence cleaved	Methylated sequence[b] not cleaved
*Eco*R I	G↓AATTC	?	GA(A)TTC
Mbo I	↓GATC	?	G(A)TC
*Sau*3A I	↓GATC	G(A)TC	GAT(C)
Taq I	T↓CGA	T(C)GA	TCG(A)
*Eco*R II	↓CCXGG	?	C(C)XGG
*Bst*N I	CC↓XGG	C(C)XGG	?

Notes: a. Recognition sequences are written from 5' → 3', only one strand being given. The mode of cleavage is indicated by the arrows. X = A or T.
b. Methylated bases have been encircled.

against cleavage. When the methylase is available in purified form, protection of recognition sites on purified DNA fragments can be obtained *in vitro.*[7]

In vitro *Recombination Using Restriction Endonucleases*

Introduction. In combination with small plasmids, restriction endonucleases can be used for the isolation of genes (i.e. for molecular cloning). A number of plasmids have been specially constructed for this purpose. These plasmids are referred to as cloning vehicles or vectors. Vectors generally have two or more selectable markers (e.g. antibiotic resistance genes) and a number of unique cleavages sites for restriction endonucleases in which foreign DNA can be inserted without affecting the stable maintenance of the vector in bacteria.

For cloning, the DNA containing the relevant genes (e.g. the chromosome of *E. coli*) is cleaved with a restriction endonuclease which creates single stranded extensions (Figure 11.1). The vector DNA is cleaved with the same restriction endonuclease. The vector DNA and chromosome DNA fragments can now associate by means of base pairing between their single stranded complementary ends. These base paired ends are a substrate for T4 DNA ligase which forms a phosphodiester linkage between the termini of the DNA fragments, linking them covalently (the covalent linkage of DNA fragments by T4 DNA ligase is often referred to as ligation). A mixture of linear and circular mono- and multimers is the result. Subsequently specially treated bacteria (see Chapter 12) are incubated with the mixture, and cells which have acquired the vector DNA or the vector DNA containing a passenger DNA fragment can be selected for by means of the selectable marker located on the vector DNA.

Choice of Restriction Endonuclease for Cloning. The choice of the restriction endonuclease will be largely dependent on the presence and location of cleavage sites on the DNA to be manipulated. For example, some restriction endonucleases may not be suitable because they cleave inside the gene to be cloned. Others may not be suitable because the cloning vehicle used does not contain a cleavage site for it.

When little is known about the DNA to be cloned, restriction endonucleases that cleave infrequently (i.e. recognise a hexanucleotide sequence) should be preferred to those that cleave frequently (i.e. recognise a tetranucleotide), because the chance that they will cleave within the area (gene) to be cloned is smaller.

Restriction endonucleases that create cohesive ends are more suitable than those that create blunt ends, because the joining of blunt ends by T4 DNA ligase is very inefficient (see Section 2). The choice is also determined by the nature of the cohesive ends created. DNA termini with GC rich or long single-stranded extensions are joined more efficiently than termini with AT rich or short single-stranded extensions.

Increasing the Efficiency of Cloning. One of the nuisances encountered when cloning with plasmids is that most transformants contain only the cloning vehicle, instead of the cloning vehicle with an insert. This happens because during ligation the cloning vehicle can recirculise by itself. Because of its smaller size, the recircularised vehicle transforms more efficiently than the recircularised cloning vehicles containing an insert.

One way to prevent recircularisation of the vehicle by itself is by removal of its terminal phosphate groups with alkaline phosphatase.[8] (Alkaline phosphatase is strongly inhibited by low concentrations of phosphate and it is advisable to dialyse the DNA before alkaline phosphate treatment – ethanol precipitation may not be adequate.) The resulting ends cannot be rejoined by T4 DNA ligase because dephosphorylated 5'-ends are not substrates for it. The phosphorylated 5'-ends of the DNA to be cloned can be joined to the 3'-hydroxyl-ends of the vehicle, thus permitting the formation of plasmid-passenger nicked circles. (A single-stranded break in double-stranded DNA is called a nick.) Since linear DNA transforms very inefficiently the background of unmodified cloning vehicle is decreased substantially.

Another way to reduce the background of recircularised cloning vehicle is by using two restriction endonucleases that have different recognition sequences, but generate the same single-stranded extensions (Table 11.3). For example, the cloning vehicle could be cleaved with *Bam*H I, while the DNA to be cloned is cleaved with *Bcl* I. After ligation, those molecules that are composed of the cloning vehicle and passenger DNA will be resistant to *Bam*H I (and *Bcl* I) cleavage, because a hybrid sequence is formed at the junctions of the two DNA fragments. If the mixture is treated with *Bam*H I after ligation, the circular cloning vehicle molecules that do not have an

Figure 11.1: Steps in Gene Cloning

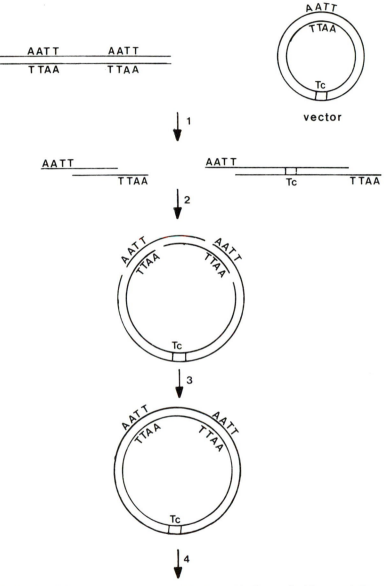

Note: 1. DNA containing the gene to be cloned is digested with a restriction endonuclease creating single-stranded extensions. The vector used for cloning is cleaved with the same restriction endonuclease. 2. Because vector DNA and the DNA fragments to be cloned now have single-stranded complementary termini, they are readily joined by base pairing between these termini. 3. T4 DNA ligase is used to stabilise these linkages by means of formation of a covelent phosphodiester bond. 4. The resulting recombinant plasmids are introduced into bacteria by transformation. Bacteria containing vector DNA can be selected for by their tetracycline resistance. Tc, tetracycline resistance genes.

insert (and thus have retained their *Bam*H I site) will be cleaved. The resulting linear DNA molecules have a very low transforming efficiency. If necessary, the cloned DNA can be excised from the cloning vehicle by (partial) cleavage with *Sau*3A I (Table 11.3).

The same principle can be applied to restriction endonucleases that generate blunt ends. Since all blunt ends can be joined with T4 DNA ligase (albeit inefficiently), any combination of restriction endonucleases that generate blunt ends can be used.

The efficiency of cloning can also be enhanced by the use of terminal deoxynucleotidyl transferase.[10] Terminal deoxynucleotidyl transferase adds a homopolymer extension to the 3'-end of DNA. If a homopolymer of dG is added to the 3'-ends of the cloning vehicle and a complementary homopolymer tail (dC) is added to the 3'-ends of the DNA to be cloned, the cloning vehicle can only recircularise with the aid of passenger DNA. By choosing an appropriate combination of homopolymer extension and restriction endonuclease, the restriction endonuclease site can be reconstituted (for example, addition of a dG extension to *Pst* I cleaved DNA, restores the *Pst* I site).

Joining of Ends Created by Different Restriction Endonucleases. As mentioned above, some restriction endonucleases generate the same single-stranded extension, although they recognise different sequences (Table 11.3) and these ends can be joined with T4 DNA ligase. Also all blunt ends can be joined by T4 DNA ligase, irrespective of the restriction endonuclease used to generate them.

It is not possible to join DNA molecules with different single-stranded extensions or with a single-stranded extension and a blunt end with T4 DNA ligase. These ends can be joined after 'filling in' or removing the single-stranded extensions to create blunt ends. A 3'-recessed end can be filled in by T4 DNA polymerase or the Klenow fragment of DNA polymerase I (a proteolytic fragment containing only the polymerase and $3' \rightarrow 5'$ exonuclease activities of the enzyme) and deoxynucleotide tri-phosphates.[11,12] A protruding 3'-end can be removed by single-strand specific exo- or endonucleases. T4 DNA polymerase can also be used for this purpose,[13] because in addition to its $5' \rightarrow 3'$ polymerase activity, this enzyme has a $3' \rightarrow 5'$ exo-nuclease activity. S1 nuclease, a specific single-stranded endonuclease,[14] can be used to remove both 3'- and 5'-single-stranded extensions. Subsequent treatment of S1 nuclease digested DNA with T4 polymerase increases the efficiency of blunt end ligation about ten-fold, presumably because most ends still have small extensions after S1 nuclease treatment.[15]

By appropriate choice of restriction endonuclease, the joining of (filled-in) ends generated by different restriction endonucleases can result in the reconstitution of a recognition site (Table 11.4).

Table 11.3: Restriction Endonucleases that have Different Recognition Sequences but Create the same Single-stranded Extensions

Restriction endonuclease	Recognition[a] sequence	Single stranded extension
*Bam*H I	G̣GATCC	GATC
Bcl I	ṬGATCA	GATC
Bgl II	ẠGATCT	GATC
*Sau*3A I	̣GATC	GATC
Cla I	AṬCGAT	CG
Asu II	TṬCGAA	CG
Taq I	ṬCGA	CG
Sal I	G̣TCGAC	TCGA
Xho I	C̣TCGAG	TCGA
*Eco*R I	G̣AATTC	AATT
*Eco*R I*	̣AATT	AATT

Note: a. Recognition sequences are written from 5′ → 3′, only one strand being given. The mode of cleavage is indicated by the arrows.

Table 11.4: Some Examples of Restriction Endonucleases that Generate Ends that may be Filled in with T4 DNA Polymerase and whose Sites can be Restored by Ligation to an Appropriate Blunt End[16]

Restriction endonuclease	Recognition sequence[a]	Ends after filling in	Requirement on blunt ends[b]
*Eco*R I	5′-G̣AATTC-3′	-GAATT-3′	5′-C-
	3′-CTTAAG̣-5′	-CTTAA-5′	3′-G-
*Bam*H I	5′-G̣GATCC-3′	-GGATC-3′	5′-C-
	3′-CCTAGG̣-5′	-CCTAG-5′	3′-G-
Sal I	5′-G̣TCGAC-3′	-GTCGA-3′	5′-C-
	3′-CAGCTG̣-5′	-CAGCT-5′	3′-G-
Hind III	5′-ẠAGCTT-3′	-AAGCT-3′	5′-T-
	3′-TTCGAẠ-5′	-TTCGA-5′	3′-A-

Notes: a. The mode of cleavage is indicated by the arrows.
    b. Created directly by cleavage with restriction endonuclease, or after filling in recessed ends or removing single-stranded extensions.

Physical Mapping of Restriction Endonuclease Cleavage Sites

Introduction. The use of restriction endonucleases is facilitated by the availability of a physical map of the DNA to be analysed. A physical map is derived by determining the location of a number of restriction endonuclease cleavage sites on the DNA relative to each other.

The strategy followed for physical mapping depends largely on the size of the DNA (plasmid) and the number of cleavage sites of the used restriction endonucleases. When studying very large plasmids for example (60 megadalton and larger) it is convenient to clone segments of the plasmid in a multicopy cloning vehicle. Mapping is simplified with these recombinant plasmids because they are less complex and a more convenient source of DNA (fragments). Moreover, mapping of DNA fragments contained in a cloning vehicle is facilitated by the availability of detailed physical maps of these plasmids.

For physical mapping it is necessary to determine the molecular weights of the (partial) cleavage products of the restriction endonucleases. Lambda DNA cleaved with *Hind* III and/or *Eco*R I is a convenient source for marker molecules.[17] For the smaller molecular weights, DNA fragments of pBR322 can be used as molecular weight markers (see Chapter 14).[18]

Especially when (partially cleaved) DNA fragments have to be separated, isolated and cleaved for a second time, it may be necessary to use ^{32}P-labelled DNA. This allows the detection of very low amounts of DNA. DNA can be labelled uniformly *in vitro* with ^{32}P by nick-translation (see Chapter 8).[19] The 5'-termini of DNA can be labelled *in vitro* with polynucleotide kinase.[20]

Mapping Cleavage Sites. If no cleavage sites are known on a plasmid, it is best to start with two or more restriction endonucleases that cleave the DNA only a few times. Restriction endonucleases that cleave the DNA only once are especially useful at this stage. A physical map can then be deduced by comparing the cleavage patterns of single digestions (enzyme A or B) with those of double digestions (enzyme A and B) (see Figure 11.2). In this connection it should be noted that the mobility of DNA fragments is influenced by the DNA and salt concentration. When comparing cleavage patterns on a gel, these factors should not differ too much between different samples.

When the positions of a few cleavage sites have been determined, the mapping of new sites is simplified. Their positions can be mapped relative to the known sites by single or double digestions, or by isolating DNA fragments generated by one (or two) restriction endonuclease(s) and cleaving them with a second (third) restriction endonuclease (Figure 11.2).

If restriction endonucleases that cleave the DNA frequently are used, other approaches are necessary. These approaches involve the determination of the fragment order by the characterisation of partially

Figure 11.2: Mapping of Cleavage Sites by Single and Double Digestions

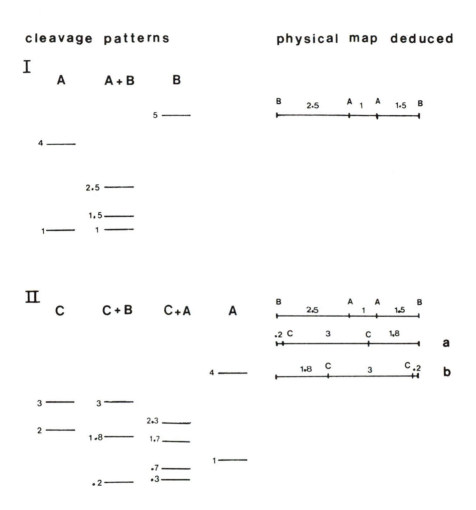

Note: I. The sites for the restriction endonucleases A and B are mapped relative to each other by determining the molecular weights of the DNA fragments generated by A, A + B and B. The physical map has been depicted in a linear form, because this is more convenient when trying to fit in the various DNA fragments. II. The sites of a third restriction endonuclease (C) are mapped relative to A and B. From the comparison of the cleavage patterns of C and C+B it is clear that B cleaves within the 2 Md C fragment. Two physical maps (a and b) can be drawn in which this is the case. From the comparison of the cleavage patterns of C+A and A it is obvious that C cleaves within the 1 Md A fragment, and thus that map b is not correct. The numbers in the cleavage patterns and in the physical maps refer to the molecular weights of the DNA fragments ($\times$ 10^6). Md = mega dalton.

cleaved DNA molecules. A plasmid is cleaved partially with a restriction endonuclease and the cleavage products are separated by means of gel electrophoresis. After determining their molecular weights, the fragments are eluted from the gel (see Chapter 16). Their composition is determined by cleaving them completely with the same restriction endonuclease and analysing the cleavage products on a gel. A disadvantage of this method is that it can be very laborious if the possible number of partially cleaved molecules is very large.

Figure 11.3: Determination of the Fragment Order by Partial Cleavage of Terminally Labelled DNA[21]

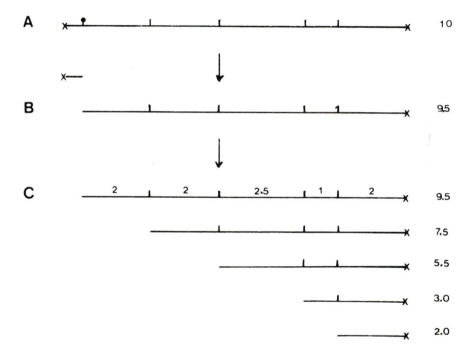

Note: A. Plasmid DNA is linearised with a restriction endonuclease. The 5′-termini of the linear DNA are dephosphorylated with alkaline phosphatase and subsequently terminally labelled with ^{32}P with polynucleotide kinase. The labelled ends have been indicated by X. B. One labelled end is removed by cleavage with a second restriction endonuclease. When the two resulting DNA fragments both contain the sites that are to be mapped, it is necessary to isolate them and treat them separately. In the given example this is not necessary. C. The labelled DNA fragment is partially cleaved with the restriction endonuclease whose sites are to be mapped. The cleavage products are separated on a gel and visualised by means of autoradiography, so that only fragments containing the labelled terminus are detected. By determining the molecular weight of these partially cleaved DNA molecules, the fragment order can be deduced. The differences in molecular weight should correspond to the molecular weights of the fragments obtained after a complete digest. The numbers refer to the molecular weights of the DNA fragments ($\times$ 10^6). The cleavage sites of the restriction endonucleases are indicated by ❢ and I.

A very elegant way to determine the order of DNA fragments is based on the same principle as DNA sequencing. Plasmid DNA is cleaved at a suitable (and preferably unique) site and the 5' termini of the linear DNA are labelled with ^{32}P.[21] If a partial digest of this DNA is analysed on a gel and visualised by means of autoradiography, only those bands that contain a labelled terminus will be seen. By determining the molecular weights of these bands it is possible to determine the fragment order. This is illustrated in Figure 11.3.

Mutagenesis in vitro *using Restriction Endonucleases*

Introduction.   After cloning it is often necessary to locate the genes contained within the cloned DNA. Also it might be important to know how the genes are organised in operons and what functions they code for. Most of these questions can be answered by the construction and analysis of mutants.

One way to obtain mutations in genes located on a cloned DNA fragment is by means of transposons.[22] However, insertions of transposons in an operon generally affect the expression of more than one gene, because transcription is terminated within the transposon. Thus, due to the pleiotropic effects of the insertion mutations it is difficult to interpret the changes in phenotype observed and to ascribe an effect to insertional inactivation of a particular gene. Furthermore, analysis of insertion mutants in minicells (see Chapter 13) is often complicated by the presence of polypeptides encoded by the transposon.

Mutants constructed *in vitro* using restriction endonucleases generally do not have these drawbacks. Moreover restriction endonucleases can be used to introduce mutations at predetermined sites. Therefore restriction endonucleases (in combination with other endo- and exonucleases) are an attractive alternative to transposons to obtain mutants.

The methods of *in vitro* mutagenesis that allow the introduction of point-mutations at predetermined sites[24,25,26] will not be discussed here. These methods are generally used for a more refined analysis, after the various genes have been located (and often sequenced) and the functions of their products determined.

Deletion of DNA Fragments.   The introduction of relatively large deletions is particularly useful when large DNA segments are to be analysed and approximate locations of specific genes and signals are to be determined. On the other hand the effect of large deletions can be complex because more than one gene can be deleted and this method is therefore less suitable to determine functions of gene products.

Some deletions can be introduced simply by cleaving the plasmid DNA with two different restriction endonucleases that have unique sites close to each other. If different staggered ends are created, these can be filled in or removed (see Section 1, 'The joining of ends created by different restriction endonucleases'). Of course if a restriction endonuclease has two or more sites in a small DNA segment, this segment can be deleted by cleavage with the restriction endonuclease, followed by ligation.

A method which allows the isolation of a whole set of deletion mutants, involves partial cleavage with a restriction endonuclease. Ideally, the restriction endonuclease should cleave the cloned DNA frequently and the cloning vehicle infrequently. The optimal conditions for partial cleavage (whereby most plasmid molecules are cleaved at two sites) can be determined empirically by either varying the enzyme concentration or the time of incubation. The reaction can be stopped by the addition of excess EDTA, SDS or by heating for ten minutes at 65°C (some restriction endonucleases (*Taq* I, *Bst*N I) are not inactivated at this temperature). Several problems may arise if the cleaved DNA is subsequently ligated and used for transformation. The uncleaved DNA molecules present will give a large background of unmodified plasmid DNA. Moreover, because small molecules ligate and transform more efficiently than large molecules, most plasmids isolated will be small and thus will contain large deletions. Finally, it is possible that small DNA fragments are rejoined to the larger molecules. The rearrangement of small fragments makes it difficult to interpret the effect of the mutation because it can be caused by rearrangement, deletion or both. Moreover, rearrangements are not always easily detected and it can be quite difficult to determine the nature of the rearrangement. Therefore it is advisable to separate those DNA molecules that have the desired deletions from uncleaved plasmid molecules and small DNA fragments by preparative gel electrophoresis. After elution from the gel (see Chapter 16) this DNA is ligated and used to transform competent cells.

It is possible to select for certain deletions if the DNA segment to be deleted contains a unique cleavage site for a restriction endonuclease. After partial cleavage and ligation the DNA is cleaved with the restriction enzyme that has the unique site in the segment to be deleted. This treatment linearises all the DNA molecules that have retained this segment.

Transformation with Linear DNA.   The main advantage of this method[27,28] is its simplicity. Plasmid DNA is linearised by means of partial cleavage or with a restriction endonuclease that cleaves only once. In the first case the linear full length molecules must be isolated by preparative gel electrophoresis before transformation. The DNA is not recircularised with T4 DNA ligase, but used directly to transform competent cells. This results in deletions of varying size starting from both ends of the linear molecules used for transformation.

A disadvantage of this method is that it is relatively time consuming to determine the extent of the deletion (e.g. by heteroduplex formation).

Introducing Small Deletions or Insertions at Restriction Endonuclease Recognition Sites. With the method which will be described below, it is possible to obtain a set of mutants with small deletions or insertions located at different recognition sites of a particular restriction endonuclease. By choice of the restriction endonuclease one can determine whether the average distance between two deletions will be large or small. Ideally, the used restriction endonucleases should cleave the cloning vehicle infrequently and the cloned DNA frequently. However, a detailed knowledge of the positions of the recognition sites of the restriction endonuclease used is not a prerequisite. A map of the recognition sites and the location of the insertions or deletions can be deduced from the analysis of the mutants, because the mutation results in the elimination of a recognition site. The principle of the method is illustrated in Figure 11.4.

In the first step, plasmid DNA is cleaved with a restriction endonuclease under conditions whereby most DNA molecules are cleaved only once. This can be achieved by performing the incubation in the presence of ethidium bromide.[29] The intercalation of ethidium bromide into DNA inhibits cleavage by restriction endonucleases. Linear DNA molecules take up more ethidium bromide than covalently closed circular DNA molecules, and are therefore more resistant to cleavage in the presence of ethidium bromide. The optimal concentration of ethidium bromide required to limit the cleavage of covalently closed circular DNA by a restriction endonuclease can be determined empirically on a small scale. Subsequently the reaction can be scaled up for preparative purposes. Especially when it is difficult to separate the linear full-length molecules from smaller partially cleaved DNA molecules it is important to control the reaction so that no such smaller molecules are formed. Otherwise a large fraction of the mutants isolated will contain large deletions resulting from the excision of one or more DNA fragments.

After cleavage with the restriction endonuclease, the cleavage products (ideally consisting mainly of covalently closed circular, open (nicked) circular and linear full-length DNA molecules) are separated on a preparative gel. The gel system should be chosen so that the linear full-length molecules migrate faster than the covelently closed circular and open circular DNA molecules. Otherwise a large background of unmodified plasmid DNA will be obtained after transformation, due to trailing of the covalently closed circular and open circular DNA molecules. The linear full-length molecules are eluted from the gel (see Chapter 16) and can be treated in a number of ways. If a restriction endonuclease which creates single-stranded extensions was used, these can be filled in or removed (see Section 1, 'Joining of ends created by different restriction endonucleases'). If the DNA molecules have blunt ends, small deletions can be introduced with Bal-31, an exonuclease that degrades the 3'- and 5'-termini of duplex DNA.[30,31] Subsequently the DNA molecules are ligated and used to transform competent cells. These manipulations result in a small insertion

Figure 11.4: Principle of the Method used to Construct Mutants with Small Insertions or Deletions at Restriction Endonuclease Cleavage Sites

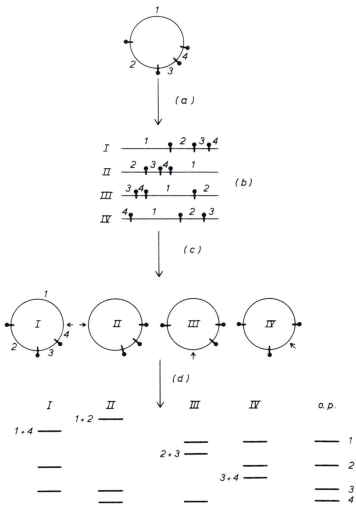

Note: a. Plasmid DNA is cleaved with a restriction enzyme in the presence of the ethidium bromide. b. The reaction products are separated on a preparative gel, and linear full-length molecules are extracted from the gel. The four possible permuted molecules have been depicted. c. Small deletions or insertions are introduced by treating the DNA ends with nuclease or polymerase. Subsequently, the DNA is circularised with T4 DNA ligase and used to transform competent cells. d. Plasmid DNA is isolated from the transformants and cleaved with the restriction enzyme. From the cleavage patterns the order of the DNA fragments and the location of the insertions or deletions can be deduced. The cleavage patterns of the four possible insertion or deletion derivatives and of the original plasmid (o.p.) are shown. The DNA fragments generated by the restriction enzyme are indicated by the numbers 1 to 4. The cleavage sites of the restriction enzyme are indicated by ↑, an eliminated cleavage site by ↓.

Source: Mooi *et al.*, *J. Bacteriol.*, 150 (1982), 512-21

or deletion at a particular restriction endonuclease recognition site, eliminating this site.

Small insertions can also be introduced by blunt-end ligation of linkers (oligonucleotides with a defined sequence containing the recognition sequence of one or more restriction endonucleases) to the termini of the linearised DNA molecules.[32] Subsequently, the DNA is cleaved with a restriction endonuclease whose recognition sequence is carried by the linker and circularised with T4 DNA ligase. Preferably a linker is chosen that carries the recognition sequence of a restriction endonuclease that does not cleave the linearised DNA molecules. If this is not possible, the internal cleavage sites can be protected by methylation *in vitro* or *in vivo*. The location of the insertion can be determined rapidly because it contains a new cleavage site.

If a DNA fragment is inserted which codes for an antibiotic resistance, the insertion can be selected for. Such fragments are conveniently isolated from the various cloning vehicles. It should be noted however, that these insertions are large and can exert polar effects within an operon.

Changing the Specificity of Substrate Recognition of a Restriction Endonuclease

For a number of restriction endonucleases it has been observed that the substrate specificity can be relaxed by changing the pH, the ionic strength, the nature of the co-factor or by the addition of an organic solvent.[33,34,35,36]

The best studied example is *Eco*R I[37] which normally recognises the hexanucleotide 5'-GAATTC-3'. At low salt concentrations and high pH, however, the tetranucleotide AATT is also cleaved (this is referred to as the *Eco*R I* activity). Fragments generated by *Eco*R I* can be cloned into *Eco*R I sites of a cloning vehicle, because the same single-stranded extensions are created. Generally, this will result in the loss of the *Eco*R I recognition site.

When preparing an incubation mixture with low salt, one should also take into account the amount of salt added with the restriction endonuclease, because they are frequently stored in solutions containing high salt.

Practical Notes

The catalytic activity of commercially available restriction endonucleases is expressed in units, whereby one unit is defined as the amount of enzyme required to completely cleave 1 μg of λ DNA under standard conditions in one hour. The amount of enzyme units (a) necessary to completely digest x μg of a plasmid (p) in one hour can be estimated with the equation

$$ a = x \cdot \frac{mw_\lambda}{\text{no. of cuts in } \lambda} \cdot \frac{\text{no. of cuts in p}}{mw_p} $$

(mw = molecular weight)
$mw_\lambda = 32$ megadalton.

Information about the number of cuts in λ can be found in Roberts.[38]

Other factors such as the base composition adjacent to the recognition site and tertiary structure of the DNA may also influence the amount of enzyme required to cleave DNA.

Although the diversity in restriction endonucleases is large, most incubations can be performed in a Tris-HCl buffer containing a low (6 mM), intermediate (50 mM), or high (150 mM) salt concentration:

> *Incubation mix*
>
> 6 mM Tris HCl pH 7.5
> 6 mM MgCl$_2$ $\Big\}$ + 6 mM, 50 mM or 150 mM NaCl.
> 6 mM B-mercaptoethanol

DNAse-free BSA (to 100 μg/ml) or gelatine can be added to the incubation mix to stabilise the enzyme and to avoid absorption of the enzyme to the vial.

If DNA is to be cleaved by more than one restriction endonuclease, mutual inhibition may result if the enzymes are added simultaneously. This can be prevented if the first enzyme is inactivated by heat treatment (10 min, 65°C) or phenol extraction prior to addition of the second enzyme.

2. The Use of T4 DNA Ligase

Introduction

T4 DNA ligase catalyses the synthesis of a phosphodiester bond between the 3'-hydroxyl and 5'-phosphoryl ends of DNA, and requires Mg^{2+} and ATP as co-factors. T4 DNA ligase is the product of gene 30 of the T4 phage. The reaction it catalyses occurs in three steps. In the first step the adenylyl group of ATP is transferred to the ε-NH$_2$ of a lysine residue in the enzyme. In the second step, the adenylyl group is transferred from the enzyme to the 5'-phosphoryl end of the DNA. In the third step, a γ phosphodiester bond is formed between a 3'-hydroxyl end and the adenylated 5'-phosphoryl end with release of AMP.

A 5'-hydroxyl end is not a substrate for T4 DNA ligase, and this fact can be used to prevent the joining of ends between particular DNA molecules by removing the terminal 5'-phosphate (see Section 1, 'Increasing the efficiency of cloning').

Various assays have been used to define the activity of T4 DNA ligase preparations. For *in vitro* recombination the most practical assay involves the determination of the amount of enzyme required to give 50% ligation of *Hin*d III fragments of λ under standard conditions.

The Joining of Cohesive or Blunt Ends

The usefulness of T4 DNA ligase for *in vitro* recombination is greatly enhanced by its ability to join blunt ends,[39] as well as cohesive ends. Although the optimal temperature of T4 DNA ligase is expected to be 37°C, much lower temperatures are used to join cohesive ends, because the short hydrogen-bonded termini, which are its substrate, are more stable at lower temperatures. The temperature for melting such ends (tm) depends on the nature of the cohesive ends, the DNA concentration and the length of the DNA fragments.[40] In general, the higher the DNA concentration, the smaller the DNA fragments and the stronger the cohesive ends the higher the temperature that can be used for the ligation. Cohesive ends are generally ligated at temperatures of 10–16°C.

The joining of blunt ends is not facilitated by alignment due to base pairing and these ends are joined with a much lower efficiency than cohesive ends. Especially at low T4 DNA ligase concentrations, blunt end joining can be stimulated by T4 RNA ligase.[41] The mechanism of this stimulation is not known. Ligation of blunt ends is generally performed at temperatures of 10–20°C.

Because the chance that two DNA ends meet and remain aligned long enough to be joined by T4 DNA ligase is small, long incubations (six to 16 hours) are necessary for optimal ligation.

When a mixture of DNA fragments with blunt and cohesive ends is used, it is possible to join selectively only the cohesive ends by using *E. coli* DNA ligase, because this enzyme cannot join blunt ends. Alternatively, T4 DNA ligase can be used in combination with high (5 mM) ATP concentrations.[42] At ATP concentrations of 5 mM, blunt end ligation by T4 DNA ligase is completely inhibited, while cohesive end ligation is only inhibited to a limited extent (8%). At higher ATP concentrations (7.5 mM) the cohesive end joining is also completely inhibited. The optimal ATP concentration for blunt end and cohesive end ligation is 0.6 mM.[43]

Intra- or Intermolecular Joining

T4 DNA ligase can catalyse intramolecular joining of ends (resulting in cyclic monomers), intermolecular joining of ends (resulting in linear multimers) or both (resulting in circular multimers). Dugaiczyk *et al.*[44] have studied the parameters which determine the extent of intra-and intermolecular joining. The extent in which these two reactions take place depend on the ratio of j : i, whereby j represents the effective concentration of one end in the neighbourhood of the other end of the same molecule, and i, the total concentration of ends. j is inversely related to the length (molecular weight) of the DNA molecules and independent of the DNA concentration, while i is dependent on the DNA concentration. The relationship between these four parameters is given by the following equation which is derived for DNA fragments having *identical* self complementary ends or blunt ends:

1. $j/i = \dfrac{51.1}{[DNA] \cdot \sqrt{mw}}$

or,

2. $[DNA] = \dfrac{51.1}{j/i \cdot \sqrt{mw}}$

mw = molecular weight
[DNA] in mg/ml

With this equation it is possible to choose conditions which preferentially promote intra- or intermolecular joining of ends. When $j/i < 1$ the formation of intermolecular bands is favoured, when $j/i > 1$, more intramolecular bonds should be formed. In principle, equal amounts of intra- and intermolecular bonds should be formed when $j/i = 1$. In practice however, significant amounts of intramolecular bonds are only formed at higher $j:i$ ratios. It should be noted that the $j:i$ ratio for a particular class of DNA fragments will increase during the course of the ligation, because j will remain constant for these fragments, while i (the concentration of ends) decreases due to the joining of termini. Thus in the course of the reaction there is a tendency towards circularisation.

If only circularisation or linearisation of DNA fragments with a particular molecular weight is desired, equation 2 can be used to calculate the required DNA concentration. However, generally the situation is more complex. For example, when cloning DNA fragments with a cloning vehicle, intermolecular joining should occur between the DNA fragments to be cloned and the cloning vehicle, followed by intramolecular joining of the resulting hybrid DNA molecules. Moreover, the DNA fragments to be cloned generally have diverse molecular weights. In these instances approximate values can be calculated as follows:

3. $j = \dfrac{615 \cdot 10^{20}}{(mw_v + mw_f)^{3/2}}$ (ends/ml)

mw_v = molecular weight cloning vehicle
mw_f = (average) molecular weight of fragments to be cloned.

4. $i = 12 \cdot 10^{20} \cdot M$ (ends/ml)
      M = molar concentration of DNA molecules

The equation is derived for DNA fragments having identical self-complementary ends or blunt ends.

M can be calculated for the desired j : i ratio.
For j/i = n (j = n.i.):

$$M = \frac{51}{n \cdot (mw_v + mw_f)^{3/2}} \quad ,$$

whereby M represents the molar concentrations of all DNA molecules

Generally a molar excess of DNA fragments to be cloned is used with respect to the vector molecules, so that the vector DNA preferentially forms intermolecular bonds with the fragments to be cloned.

Further Reading

Methods in Enzymology, 'Nucleic Acids'. Eds: L. Grossman and K. Moldave (Academic Press, 1980), 65.
Methods in Enzymology, 'Recombinant DNA'. Ed: R. Wu (Academic press, 1979), 68.
A. Kornberg, 'DNA Replication' (Freeman and Co.).

References

1. R. Yuan, 'Structure and mechanism of multifunctional restriction endonucleases', *Ann. Rev. Bioch., 50* (1981), 285–315.
2. R.J. Roberts, 'Restriction and modification enzymes and their recognition sequences', *Nucleic Acids Res., 9* (1981), R75–R96.
3. Ibid.
4. M. McClelland, 'The effect of sequence specific DNA methylation on restriction endonuclease cleavage', *Nucleic Acids Res., 9* (1981), 5859–65.
5. J.G. Bogdarina, L.M. Vagabova and Y.I. Buryanov, 'DNA cytosine methylation in *E. coli* MRE 600 cells', *FEBS Lett., 68* (1976), 177–80.
6. McClelland, 'The effect of sequence specific DNA methylation'.
7. F. Heffron, M. So and B.J. McCarthy, '*In vitro* mutagenesis of a circular DNA molecule using synthetic restriction sites', *Proc. Natl. Acad. Sci. USA, 75* (1978), 6012–16.
8. A. Ullrich, J. Shine, J. Chirgwin, R. Pictet, E. Tischer, W.J. Rutter and H.M. Goodman, 'Rat insulin genes: Construction of plasmids containing the coding sequences', *Science, 196* (1977), 1313–19.
9. Ibid.
10. R. Roychoudbury, E. Jay and R. Wu, 'Terminal labeling and addition of homopolymer tracts to duplex DNA fragments by terminal deoxynucleotidyl transferase', *Nucleic Acids Res., 3* (1976), 863–77.
11. R.M. Wartell and W.S. Reznikoff, 'Cloning of DNA restriction endonuclease fragments with protruding single-stranded ends', *Gene, 9* (1980), 307–19.
12. K. Backman, M. Ptashne and W. Gilbert, 'Construction of plasmids carrying the CI gene of bacteriophage λ', *Proc. Natl. Acad. Sci. USA, 73* (1976), 4174–78.
13. Wartell and Reznikoff, 'Cloning of DNA restriction endonuclease fragments'.
14. R.C. Wiegand, C.N. Godron and C.M. Radding, 'Specificity of the S1 nuclease from *Aspergillus oryzae*', *J. Biol. Chem., 250* (1975), 8848–55.
15. K. Shishido and T. Ando, 'Efficiency of T4 DNA ligase catalyzed end joining after S1 endonuclease treatment on duplex DNA containing single-stranded portions', *Biochim. Biophys. Acta, 656* (1981), 123–7.
16. Blackman, Ptashne and Gilbert, 'Construction of plasmids'.
17. K. Murray and N.E. Murray, 'Phage lambda receptor chromosomes for DNA fragments made with restriction endonuclease III of *Haemophilus influenzae* and restriction endonuclease I of *Escherichia coli*', *J. Mol. Biol., 98* (1975), 551–64.

18. J.G. Sutcliffe, 'pBR322 restriction map derived from the DNA sequence: accurate DNA size markers up to 4361 nucleotide pairs long', *Nucleic Acids Res., 5* (1978), 2721–8.
19. T. Maniatis, A. Jeffrey and D.G. Kleid, 'Nucleotide sequence of the rightward operator of phage λ', *Proc. Natl. Acad. Sci. USA, 72* (1975), 1184–8.
20. A.M. Maxam and W. Gilbert, 'A new method for sequencing DNA', *Proc. Natl. Acad. Sci. USA, 74* (1977), 560–4.
21. H.O. Smith and M.L. Birnstiel, 'A simple method for DNA restriction site mapping', *Nucleic Acids Res., 3* (1976), 2387–98.
22. G. Dougan and D. Sherratt, 'The transposon *Tn* 1 as a probe for studying Col E1 structure and function', *Mol. Gen. Genet., 144* (1977), 243–51.
23. Ibid.
24. W. Müller, H. Weber, F. Meyer and C. Weismann, 'Site directed mutagenesis in DNA; Generation of point mutations in cloned 3-globulin complementary DNA at positions corresponding to amino acids 121 to 123', *J. Mol. Biol., 124* (1978), 343–58.
25. C.A. Hutchinson, S. Phillips, M.H. Edgell, S. Gillam, P. Jahnke and M. Smith, 'Mutagenesis at a specific position in a DNA sequence', *J. Biol. Chem., 255* (1978), 6551–60.
26. D. Shortle, D. Koshland, G.M. Weinstock and D. Bolstein, 'Segment directed mutagenesis: Construction *in vitro* of point mutations limited to a small predetermined region of a circular DNA molecule', *Proc. Natl. Acad. Sci. USA, 77* (1980), 5375–79.
27. R. Thompson and M. Achtman, 'The control region of the F sex factor DNA transfer cistrons: physical mapping by deletion analysis', *Mol. Gen. Genet., 169* (1979), 49–57.
28. C. Lai and J. Nathans, 'Deletion mutants of semian virus 40 generated by the enzymatic excision of DNA segments from the viral genome', *J. Mol. Biol., 89* (1974), 179–83.
29. R.C. Parker, R.M. Watson and J. Vinograd, 'Mapping of closed circular DNAs by cleavage with restriction endonucleases and calibration by agarose gel electrophoresis', *Proc. Natl. Acad. USA, 74* (1977), 851–5.
30. H.B. Gray, D.A. Ostrander, J.L. Hodnett, R.J. Legerski and D.L. Robberson, 'Extracellular nucleases of *Pseudomonas* Bal 31. I. Characterization of single-strand specific deoxyribonuclease and double-strand deoxyribonuclease activities', *Nucleic Acids Res., 2* (1975), 1459–92.
31. R.J. Legerski, J.L. Hodnett and H.B. Gray, 'Extracellular nucleases of *Pseudomonas* Bal 31. III. Use of the double-strand deoxyribonuclease activity as the basis of a convenient method for mapping of fragments of DNA produced by cleavage with restriction enzymes', *Nucleic Acids Res., 5* (1978), 1445–63.
32. Ullrich, Shine, Chirgwin, Pictet, Tischer, Rutter and Goodman, 'Rat insulin genes'.
33. B. Polyski, P. Green, D.E. Garfin, B.J. McCarthy, H.M. Goodman and H.W. Boyer, 'Specificity of substrate recognition by the *Eco*R I restriction endonuclease', *Proc. Natl. Acad. USA, 72* (1975), 3310–14.
34. V.A. Kolesnikov, V.V. Zinoview, L.N. Yashina, V.A. Karginov, M.M. Bachanov and E.G. Malygin, 'Relaxed specificity of endonuclease *Bam*H I as determined by identification of recognition sites in SV40 and pBR322 DNAs', *FEBS Letts., 132* (1981), 101–4.
35. J. George, R.W. Blakesley and J.G. Chirikjian, 'Sequence-specific endonuclease *Bam*H I: Effect of hydrophobic reagents on sequence recognition and catalysis', *J. Biol. Chem., 255* (1979), 6521–4.
36. E. Malyguine, P. Vannier and P. Yot, 'Alteration of the specificity of restriction endonucleases in the presence of organic solvents', *Gene, 8* (1980), 163–77.
37. Polyski, Green, Garfin, McCarthy, Goodman and Boyer, 'Relaxed specificity of endonuclease *Eco*R I'.
38. Roberts, 'Restriction and modification enzymes'.
39. V. Sgaramella, J.H. van de Sande and H.G. Khorana, 'Studies on polynucleotides, C. A novel joining reaction catalyzed by the T4-polynucleotide ligase', *Proc. Natl. Acad. Sci. USA, 67* (1975), 1468–75.
40. A. Dugaiczyk, H.W. Boyer and H.M. Goodman, 'Ligation of *Eco*I endonuclease-generated DNA fragments into linear and circular structures', *J. Mol. Biol., 96* (1975), 171–84.
41. A. Sugino, H.M. Goodman, H.L. Heyneker, J. Shine, H.W. Boyer and N.R. Cozzarelli, 'Interaction of bacteriophage T4 RNA and DNA ligases in joining of duplex DNA at base-paired ends', *J. Biol. Chem., 252* (1977), 3987–94.

42. L. Ferretti and V. Sgaramella, 'Specific and reversible inhibition of the blunt end joining activity of the T4 DNA ligase', *Nucleic Acids, 9* (1981), 3695–705.
43. Ibid.
44. Dugaiczyk, Boyer and Goodman, 'Ligation of *Eco* I'.

12 MOLECULAR CLONING OF cDNA: BACTERIAL TRANSFORMATION AND SCREENING OF TRANSFORMANTS

Brian G. Forde

CONTENTS

1. Introduction

Once a DNA fragment (or a mixture of DNA fragments) has been ligated into a suitable vector (see Chapter 11), the next step towards the cloning of that DNA fragment is the introduction of the hybrid vector DNA into an appropriate host, usually *Escherichia coli*. A cell which has acquired a heritable change in its genotype by taking up naked DNA from its environment is said to have been genetically transformed; the line of cells derived from each independently transformed cell is referred to as a 'clone'. Techniques for transforming *E. coli* are discussed in Section 2 of this chapter.

Fewer than 1 in 1000 of the cells which have been treated with the vector DNA will actually be transformed and often only a tiny fraction of those transformed cells will have received the DNA fragment of interest. It may sometimes be possible, when cloning bacterial or yeast genes, to select for cells carrying a particular gene by making use of that gene's ability to complement a mutation in the host strain. More frequently, however, the screening of a collection of clones (also known as a 'bank' or 'library') requires a step-by-step process of elimination in which progressively more discriminating (and laborious) techniques are applied as the number of potential candidates is reduced. This is particularly true in the case of libraries of double-stranded cDNA (ds-cDNA, see Chapter 9) which are often derived from complex mixtures of mRNAs in which the sequence of interest is in low abundance and for which no specific hybridization probe is available. Chapter 9 has described the construction of ds-cDNA and methods for enrichment for sequences of interest prior to cloning. In Section 3 of this chapter the techniques available for screening a cDNA library are described. Although the emphasis is on cDNA libraries, many of the same techniques could be applied to the screening of any fairly small collection of clones.

2. Transformation

Although the transformation of many bacterial genera and species can occur naturally, the species of most interest to molecular biologists, *Escherichia coli*, remained refractory to transformation until 1970. In that year, Mandel and Higa showed that cells of *E. coli* which had been treated with $CaCl_2$ could take up purified bacteriophage λ DNA, from which viable phage particles were produced.[1] Two years later Cohen *et al.* showed that similarly treated *E. coli* cells could be transformed with closed circular plasmid DNA.[2]

The procedure, as later described in detail by Lederberg and Cohen,[3] starts with the preparation of an exponentially growing *E. coli* culture. After collection of the cells by low-speed centrifugation they are carefully washed, first in ice-cold $MgCl_2$ (0.1 M), then in ice-cold $CaCl_2$ (0.1 M). After a 20 minute period of incubation in $CaCl_2$, the cells are able to take up exogenous DNA and are said to be 'competent'. At this stage the recombinant DNA is added, and after a further 30 minutes on ice the cells are given a heat pulse for two minutes at 42°C and then inoculated into nutrient broth. Usually the plasmid vector carries a drug resistance gene to allow selection for those cells which have been transformed. However the cells are not immediately brought into contact with the antibiotic because a period of incubation (90–120 minutes) at 37°C in the nutrient broth is required to allow the antibiotic resistance to be expressed. Transformation is then assayed by plating the cells out in nutrient agar plates containing the antibiotic and incubating them overnight at 37°C.

The same protocol, with minor modifications, has been widely used for the transformation of *E. coli* with recombinant plasmids. The transformation frequencies obtained however are generally low, yielding about $1-3 \times 10^6$ transformants per μg of supercoiled pBR322, but only 10^4 or 10^3 per μg of linearised and re-ligated plasmid DNA. This means that even with supercoiled DNA there is only one transformant for every 10^5 molecules of DNA. Recently, however, a simple modification to the $CaCl_2$ treatment has improved the efficiency of transformation by a factor of ten.[4] The improvement is achieved by extending the time of incubation in $CaCl_2$ from 30 minutes to 24 hours, with the result that about 20% of the cells which remain viable can be transformed. Similar efficiencies can be obtained by treating the cells with a combination of rubidium and calcium chloride, but the procedure is a more complex and laborious one.[5]

Since the preparation of competent cells, even in its simplest form, involves several steps and spans most of one working day, a procedure has been developed to allow competent cells to be stored frozen in glycerol.[6] Using this technique it is possible to keep a batch of competent *E. coli* cells at −82°C for at least 15 months without further loss of transformation efficiency. When needed, tubes of frozen cells are simply thawed in an iced water bath for 10 minutes and the DNA is added. However, the transformation efficiencies obtained with frozen cells are rather lower than those obtained with freshly prepared cells. When attempting to clone small amounts of cDNA it may therefore be preferable to prepare competent cells freshly for each experiment, despite the additional effort required.

3. Screening Transformants

Screening the transformants is normally carried out in stages, starting with techniques that can be readily applied to large numbers of clones and then, as the number of hopeful candidates is reduced, resorting to more discriminating, but laborious, techniques. As will be seen, however, the screening methods which are used will be dependent on factors such as the abundance of the sequence of interest in the ds-cDNA and the state of knowledge about the protein to which it is related.

Insertional Inactivation

As already mentioned, successful transformants are usually identified by the antibiotic resistance which they have acquired from the plasmid vector. However, not all of these transformants will harbour plasmids which contain inserted DNA. Some plasmid molecules will have remained undigested by the restriction enzyme used to linearise them, while others may have recircularised without incorporating a DNA fragment. It is therefore often useful to be able to identify those clones which do have DNA inserts. Usually this is done by testing for the inactivation of an antibiotic resistance gene. For example, pBR322 carries two antibiotic resistance markers, tetracycline resistance (tet^R) and ampicillin resistance (amp^R). Within the *tet*^{R} gene are recognition sequences for the restriction enzymes *BamH* 1 and *Hind* III. Insertion of a DNA fragment into either of these sites usually inactivates the *tet*^{R} gene. Therefore once the amp^R transformants have been selected the colonies are 'replica-plated' onto agar plates containing ampicillin. This involves transferring the pattern of bacterial growth from the initial 'master' plate onto 'replica' plates containing either ampicillin alone or tetracycline and ampicillin. This can be done by pressing a velvet pad onto the surface of the master plate and then onto each of the other two plates in succession. Alternatively, if the number of clones is small (say > 300–400), the colonies can be sampled individually using a sterile wire loop or toothpick and inoculated in an orderly array onto the secondary plates. However there is also a method which allows the positive selection of tet^S cells and obviates the need for replica plating. The transformed, amp^R cells are maintained for a period on medium containing both tetracycline and an antibiotic that will only kill growing cells. Tet^R cells grow and are killed, tet^S cells are prevented from growing and survive. The procedure is possible because, although tetracycline is bacteriostatic, there are many antibiotics, such as penicillin, which are bacteriocidal. However, penicillin is detoxified by the product of the *amp*^{R} gene. Instead, D-cycloserine is usually chosen as the bacteriocidal antibiotic for positive selection of tet^S cells.[7] Other examples are reviewed by Morrow,[7] who also discusses the occasional instances when gene inactivation has been found to be an unreliable indicator of DNA insertion.

The need for a genetic marker for DNA insertion can be greatly reduced or even eliminated by suitable treatment of the plasmid before insertion of the DNA (Chapter 11 and ref. 7). After cleavage with the restriction enzyme, for example, the terminal phosphoryl groups can be removed by phosphatase treatment and this will prevent rejoining of the plasmid without an insert. Alternatively, the use of homopolymeric tails for insertion of the cDNA will also reduce the frequency of self-ligation.

Colony Hybridization

An early step in the screening of transformants makes use of a technique devised by Grunstein and Hogness[9] for the hybridization of ^{32}P-labelled probes to bacterial colonies *in situ*. The colonies are replica-plated onto nitrocellulose filters which lie on nutrient agar plates. After an incubation period of eight to 12 hours, when the colonies have grown to an appropriate size, the filters are placed on filter paper pads which have been saturated with alkali. In this way the cells are lysed and the DNA is denatured while the colonies are still in position on the filters. After a series of washes the filters are baked under vacuum for two hours to immobilise the DNA. Once the filters have been prepared (a morning's work) they are exposed to a solution containing a ^{32}P-labelled RNA or DNA probe and hybridisation is allowed to proceed overnight. When the unhybridised RNA or DNA has been washed off, the filters are exposed to X-ray film for a few days. After development of the autoradiograph the film is matched up with the filters and those colonies which produce a dark spot on X-ray film are identified as positive.

A detailed and updated discussion of the procedures for colony hybridisation has been published by Grunstein and Wallis.[10] Modifications introduced by Thayer[11] are claimed to increase the sensitivity and reproducibility of the technique by improving the efficiency of cell lysis and immobilisation of the DNA. Recently Hanahan and Meselson[12] have modified the protocol to allow the storage and rapid screening of large numbers of colonies at high density (more than 100,000 colonies per 150 mm diameter plate). An alternative method which aids rapid screening involves the culturing of the cells in 96-well microtitre plates.[13] A home-made replica-plating device[14] can then be used for inoculating the nitrocellulose filters simultaneously with 48 or 96 clones.

The effectiveness of colony hybridisation as a screening method is largely dependent on the specificity of the radioactive probe. Ideally the probe would hybridise only to the clones of interest. Unfortunately such a probe is rarely available, unless the cDNA has already been cloned and additional or longer cDNA inserts are being sought. However, when the amino acid sequence of the protein is known it is sometimes possible to construct a synthetic oligonucleotide which will hybridise to the cDNA. Noda *et al.*[15] synthesised two oligodeoxynucleotides, each 14 nucleotides long, which represent the only two possible cDNA sequences corresponding to the sequence Glu-Trp-Trp-Met-Asp found in bovine adrenal enkephalin. These oligonucleotides were successfully used as a hybridisation probe to screen 190,000 transformants for a handful of preproenkephalin cDNA clones. Instead of using the synthetic oligonucleotide directly as a probe, Hudson *et al.* used it as a primer to synthesise a rat relaxin-specific cDNA probe from the mRNA.[16] Although synthetic oligonucleotides have been used only rarely in the past they are becoming more widely employed as the cost and availability of custom-made oligonucleotides become more favourable. The use of synthetic oligonucleotides as probes is discussed in detail by Szostak *et al.*[17]

Figure 12.1: In situ *Colony Hybridisation to a Barley cDNA Library*

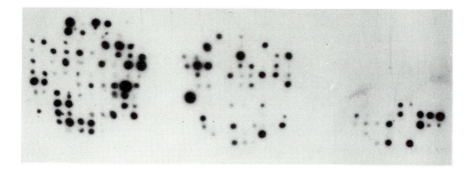

Note: The construction of the library in pBR322 is described by Forde *et al.*[19] About 270 tet-r *E. coli* transformants were replica-plated and grown overnight on three nitrocellulose filters. RNA from the same RNA fraction of barley endosperm used for cDNA cloning was ^{32}P-labelled using polynucleotide kinase. The ^{32}P-RNA was hybridised to the colonies by following the procedures of Grunstein and Hogness.[20] The autoradiograph of the filters shows that the RNA hybridised to about 130 of the clones, which are therefore identified as carrying cDNA inserts.

Even without a specific hybridisation probe it is still possible to gain a great deal of useful information from *in situ* colony hybridisation. One probe which is frequently used is an mRNA fraction similar to the one used as the template for cDNA cloning. The RNA is ^{32}P-labelled using polynucleotide kinase[18] or ^{32}P-cDNA is prepared from it by reverse transcription. This probe will identify the majority of the clones which harbour cDNA sequences (Figure 12.1): only those which contain very small cDNA inserts, or whose cDNA inserts represent mRNAs of very low abundance, will fail to be detected. To improve the specificity of the hybridisation it is possible to enrich the mRNA for the sequence of interest by fractionating it on sucrose gradients or agarose gels or, in some circumstances, by 'hybridisation-subtraction' chromatography (see Chapter 9). If an additional colony hybridisation is then carried out using a probe which is specifically *depleted* in the desired sequence, it is often possible to identify a small number of clones which hybridise more weakly to the 'depleted' probe than to the 'enriched' probe. Adjacent RNA fractions from an agarose gel or a sucrose gradient may provide one possible source of depleted and enriched probes.[21] Alternatively RNA might be isolated from cells which are induced and from cells from are uninduced for expression of the genes of interest.[22,23]

In situ colony hybridisation is primarily designed as a technique for the rapid screening of a clone bank and has a number of limitations. Grunstein and Wallis noted, for example, that the extent of hybridisation to the immobilised colonies was not greatly altered by a 1000-fold reduction in the concentration of the RNA probe.[24] This might suggest that differences in mRNA abundance between 'depleted' and 'enriched' probes of less than a 1000-fold may not be sufficient to be detectable when comparative colony hybridisations are attempted. For a more sensitive assay it may sometimes be necessary to use the technique of 'dot' hybridisation.[25] Although similar in principle to colony hybridisations, dot hybridisations are considerably more laborious as they require the initial isolation of plasmid DNA from each clone. The plasmid may be extracted by one of the rapid 'mini' plasmid preparations[26] and the DNA is spotted onto nitrocellulose filters and hybridisation carried out in the normal way.

'Single-colony-lysate' gels

A method is available which allows a large number of clones to be rapidly analysed for the size of their plasmid DNA.[27,28] As well as identifying the clones with the largest DNA inserts the technique can be extended to allow hybridisation analysis of the plasmids. A part of each selected colony is lifted from the agar plate, suspended in a small volume of buffer containing detergent and heated for a short time. The cell lysate is then electrophoresed on an agarose slab gel for several hours (see Chapter 14). Since the plasmid DNAs are very much smaller than the bacterial chromosomal DNA they run faster on the gel and are quickly separated. When the gel is stained with ethidium bromide the DNA can be viewed and photographed by its fluorescence on a u.v. light box to identify the plasmids with the lowest electrophoretic mobilities. As many as 40 colonies can be analysed on a single gel. However, because the insert DNA is not excised from the plasmid it is not possible to make an accurate estimate of its size by this method.

After the gel has been stained and photographed the plasmid DNAs can be transferred from the gel to a nitrocellulose filter according to the method of Southern[29] and hybridised to a ^{32}P-labelled probe (see Chapters 8 and 15). The 'Southern' blotting of single-colony-lysate gels may therefore be used as an alternative to the 'dot' hybridisation method in cases where *in situ* colony hybridisations are not sufficiently sensitive or reproducible.

Figure 12.2: Diagram Illustrating Two Alternative Techniques for Determining the Coding Properties of a cDNA Clone

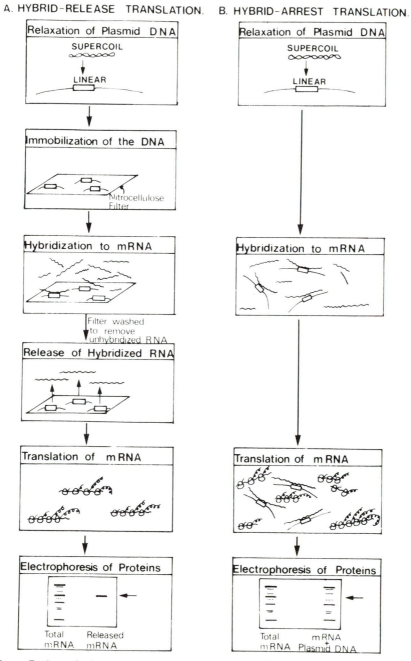

A. HYBRID-RELEASE TRANSLATION.   B. HYBRID-ARREST TRANSLATION.

Relaxation of Plasmid DNA

SUPERCOIL

LINEAR

Immobilization of the DNA

Nitrocellulose Filter

Hybridization to mRNA

Filter washed to remove unhybridized RNA

Release of Hybridized RNA

Translation of mRNA

Electrophoresis of Proteins

Total mRNA   Released mRNA

Relaxation of Plasmid DNA

SUPERCOIL

LINEAR

Hybridization to mRNA

Translation of mRNA

Electrophoresis of Proteins

Total mRNA   mRNA + Plasmid DNA

Note: Each method relies on the ability of the cloned cDNA to hybridise to the homologous mRNA. The differences between the techniques are discussed in the text.

'Hybrid-arrest' and 'Hybrid-release' Translations

Two different, but related, techniques have been devised that allow one to identify the polypeptide to which a particular cDNA clone is related (Figure 12.2). Both rely on hybridisation of the plasmid DNA to mRNA, followed by translation of the RNA in an *in vitro* protein synthesising system (see Chapter 7).

The principle on which hybrid-arrest translation (HART) is based is the inability of an mRNA-cDNA hybrid to be translated.[30] The supercoiled plasmid DNA is first relaxed (converted to open circles or linear molecules) by digestion with a restriction endonuclease which cuts it in only one or a few places, or (more economically) by nicking with S1 nuclease.[31] The relaxed plasmid is then denatured and allowed to hybridise to the mRNA fraction, after which the mRNA is used to programme the *in vitro* translation system. The polypeptide to which the cDNA insert is related is then identified by its absence from the translation products when they are analysed by gel electrophoresis and auto-radiography.

In contrast, hybrid-release translation is an assay for the ability of the plasmid DNA to select an mRNA species from the RNA population. The plasmid DNA is relaxed and denatured as before, but then it is immobilised on nitrocellulose[32,33] or diazobenzyloxymethyl (DBM) paper.[34] The immobilised DNA is allowed to hybridise to the mRNA fraction for several hours and the RNA which fails to bind is washed off. The RNA which has hybridised to the cDNA insert can be released from the filters by heating them briefly in water at 100°C. The plasmid-selected mRNA is then translated in the *in vitro* system and the products are identified by gel electrophoresis and autoradiography as before (Figure 12.3). This method has a number of advantages over that of HART. First, since it is common for two or more unrelated popypeptides to have the same electrophoretic mobility, it will often happen in HART that the lack of synthesis of a minor polypeptide will go unnoticed.[37] Secondly, a short cDNA clone may hybridise only to the 3' untranslated region of the mRNA. Although this is sufficient to allow hybrid-release translation, the mRNA-cRNA hybrid formed will be translated normally in HART and the hybridisation will fail to be detected.[38]

Identification of the 'hybrid-arrested' or 'hybrid-released' translation product as the polypeptide of interest will usually depend initially on its electrophoretic mobility. The preliminary identification may then be confirmed with the help of monospecific antibodies, which are used to immunoprecipitate the translation products.

Hybrid-release translation may also be adapted to allow the rapid screening of large numbers of clones. Derynck *et al.*[39] screened over 500 clones for interferon cDNA by pooling them in twelve groups of 46 before purifying the plasmid DNA. Each group of plasmid DNAs was then used to select mRNA and the selected mRNAs were translated in Xenopus oocytes.

*Figure 12.3: 'Hybrid-release' Translation: Fluorograph Showing the
Translation Products of mRNAs Selected by Several Different
Recombinant Plasmids*

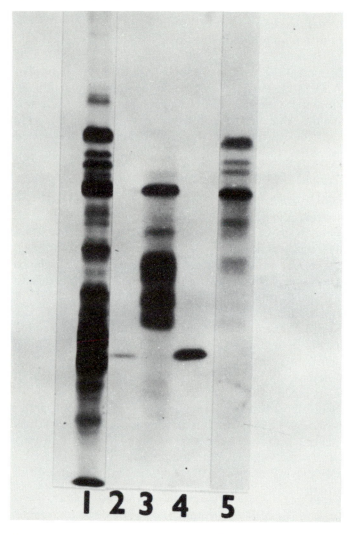

Note: DNA of each plasmid was immobilised on a nitrocellulose filter and was then
hybridised to mRNA from barley endosperm.[35] The mRNAs which hybridised were
eluted from the filters and translated in a wheat-germ cell-free protein synthesising
system, which included ^{3}H-labelled amino acids. The radioactively labelled translation
products were then analysed electrophoretically in a polyacrylamide gel containing
sodium dodecyl sulphate and the gel was fluorographed. The mRNAs used to prime
the wheat-germ system were: total polyadenylated RNA from membrane-bound
polysomes of barley endosperm; RNA selected from this fraction by pHvE-c169; RNA
selected by pHvE-c276; RNA selected by pHvE-c114 and RNA selected by pHvE-c179.
The construction of these four barley cDNA clones was described in Forde *et al.*[36]

This figure kindly provided by M. Kreis.

The production of interferon in the oocytes was detected by a sensitive bioassay. Once a group which selected interferon mRNA was identified, it was subdivided into smaller groups and the process repeated until the individual cDNA clones responsible had been located. A similar strategy would be applicable to screening for any cDNA clone of low abundance, so long as the polypeptide is readily identifiable, say by immunoprecipitation.

Even when the translation product of plasmid-selected mRNA is immunoprecipitable with antiserum this cannot be relied on as final proof of the identity of that cDNA clone. It is possible that the immunoprecipitated polypeptide is a different protein which cross-reacts with the antiserum used and happens to have the same electrophoretic mobility. This can be tested by carrying out a 'fingerprint' analysis of the immunoprecipitated polypeptide and comparing it with the authentic protein.[40]

'Northern' Hybridisation

It is sometimes useful to determine the molecular weight of the mRNA to which a cDNA clone is related. Among other things this can allow an estimate of the maximum size of the corresponding polypeptide. The RNA is first fractionated by size on a denaturing agarose gel and then transferred from the gel to DBM paper[41] or to nitrocellulose.[42] Because of its analogy to the DNA transfer method of Southern, the transfer of RNA from gels to filters is known as a 'Northern' blot. Once immobilised on the filter the RNA is hybridised to the plasmid DNA, which has been labelled with ^{32}P by 'nick translation' (Chapter 8).[43] The mRNA species which is homologous to the cloned cDNA fragment is then identified by autoradiography. If suitable molecular weight markers, such as *E. coli* ribosomal RNA, globin mRNA and tRNA, are co-electrophoresed, an estimate of the molecular weight of the hybridising mRNA can be made.

Macleod *et al.*[44] were able to use a mutant nematode strain to confirm the identity of a myosin cDNA clone by Northern hybridisation. RNA was isolated from normal cells and from a deletion mutant which synthesised a myosin protein of lower molecular weight than the wild type. It was found that the plasmid DNA hybridised to an RNA species which showed the expected reduction in size.

Screening by Detection of Bacterial Expression of the cDNA (see also Chapter 13)

Bacterial colonies can be rapidly screened with antibodies to detect the expression of foreign DNA sequences.[45,46,47] Unlike other techniques for screening colony banks for the expression of foreign DNA, *in situ* immunoassays are not dependent on the synthesis of an entire, functional protein. In theory, therefore, the technique ought to be suitable for screening cDNA clones, in which often only a fraction of the coding sequence is present. For expression to occur, the cDNA insert must be located at a site in the vector which places it under the control of a bacterial promotor and adjacent to bacterial sequences for ribosome binding and the initiation of translation.[48] In one very simple technique[49] the clones are grown overnight on agar plates containing the appropriate antiserum. The cells are lysed by overlaying the colonies with agar containing lyzosome, incubating for 1 hour and then adding a detergent solution. Incubation is continued overnight to complete the lysis and to allow immunoprecipitation of the antigen. Clones expressing the protein are detected by the appearance of immunoprecipitin rings around the colonies. A more sensitive, but more complicated, technique for *in situ* immunoassay involves the use of radioactively labelled antibodies.[50]

Perhaps the major drawback to this method for screening cDNA clones is that it will pick up less than one in six of the clones which harbour the sequence of interest. This is because of the necessity for the insert to be in the correct reading frame to encode the protein. (The ds-cDNA fragment can be inserted in two possible orientations and in each case may be translated in any of the three reading frames.) In addition, a certain proportion of those cDNA inserts which happen to be correctly orientated and positioned will synthesise protein fragments which lack the antigenic determinants. Therefore, although full-length cDNA clones are not essential for the success of this screening method, synthesis of a high proportion of full-length ds-cDNA molecules will considerably improve the chances of finding the sought-after cDNA clone. It may be these considerations which have been responsible for the failure of these techniques to find wide use as a method for screening cDNA libraries.

Restriction Enzyme Analysis

As well as providing a means of accurately estimating the size of the cDNA insert, restriction enzyme analysis (see also Chapter 11) can also be used for clone identification. There are two circumstances under which it may be possible to predict the presence of a characteristic restriction fragment in cloned cDNA sequences and then to use this information for screening purposes. In the first case a restriction enzyme analysis must be carried out on the population of ^{32}P-labelled single- or double-stranded cDNAs, as discussed in Chapter 9. Restriction fragments which are identified as being derived from the mRNA of interest can then be looked for in digests of the recombinant plasmids. For this rather time-consuming approach to be possible it is necessary for the sequence to be moderately abundant in the cDNA or ds-cDNA population ($> 2\%$).

Occasionally, when the amino acid sequence of a protein is known, it is possible to predict the location of certain restriction sites in the coding sequence of that protein. Because of the degeneracy of the genetic code it is usually not possible to predict the presence of a restriction site from the amino acid sequence. However, a few restriction enzymes have recognition sequences which break this rule. *Sau*96I recognises the sequence GGNCC (where N is any deoxynucleotide). As the four GGN codons specify glycine and the four codons specify proline, every gly-pro dipeptide will be encoded by a *Sau*96I recognition sequence. Similarly every glu-ser and asp-ser dipeptide must be encoded by a *Hinf*1 recognition sequence, and every Ala-Ala dipeptide by a *Fnu*4H1 recognition sequence. Of course, each of these recognition sequences may also occur elsewhere in the coding sequence, since each may be translated in two other reading frames to encode other amino acid sequences. Fiddes and Goodman used the amino acid sequence of the β-subunit of human chorionic gonadotropin (βHCG) to deduce the likely appearance of restriction fragments 9, 54 and 39 bp long after *Sau*96I digestion of a βHCG cDNA recombinant.[51] In this way, two out of the 72 transformants examined were identified as potential βHCG clones, and the identification was subsequently confirmed by nucleotide sequencing.

To make it easier to screen large numbers of clones, a variety of rapid procedures for isolating small amounts of relatively pure plasmid DNA have been developed, some of which are described by Davis *et al.*[52] The sensitivity of the analysis can be improved by end-labelling the restriction fragments with ^{32}P so that after electrophoresis they can be detected by autoradiography.[53] The increased sensitivity which radiolabelling provides is particularly useful for locating very small DNA fragments (< 100 base pairs).

Nucleotide Sequencing

Once a small number of clones have been selected as likely carriers of the desired cDNA fragment, their identity may be confirmed by determining their nucleotide sequence (see Chapter 16). Positive identification of a cDNA clone by its nucleotide sequence normally depends on comparison of the deduced amino acid sequence with the amino acid sequence of the protein, which may be known in full or in part. If, however, the protein has a particularly unusual amino acid composition, comparison of this with the composition of the deduced amino acid sequence may also give a good indication of the identity of the cDNA clone.

Garoff *et al.* have recently described a peptide sequencing procedure which makes it possible to verify an amino acid sequence which has been deduced from a nucleotide sequence. This technique, which requires only small amounts of the protein, should allow the identity of a cDNA clone to be confirmed even when the amino acid sequence is not known in advance.[54]

4. Conclusions

Of the wide variety of techniques for screening which are discussed in this chapter it is unlikely that more than a few will need to be applied in the search for any one cDNA fragment. Clearly the most difficult problems will be encountered in screening for a sequence of low abundance when the amino acid sequence of the protein is not known and when no specific hybridisation probe is available. Nevertheless Persico *et al.* were able to identify, under just these circumstances and using similar techniques to those outlined here, four cDNA clones related to human glucose-6-phosphate dehydrogenase from a colony bank of 3470 tet^R amp^S transformants.[55] They achieved this by judicious application of colony hybridisations, followed by the screening of 252 candidate clones in groups by hybrid-release translation. However, without knowledge of the amino acid sequence of the protein, the final, positive identification of a cDNA clone may be difficult. The problem can be overcome if a mutant is available which has a well-characterised alteration at the single genetic locus in question. One example of how such a mutant can allow a clone to be unambiguously identified by Northern hybridisation has already been given (see Section 3, 'Northern hybridisation'). An alternative method, which could be applied even when there is no mRNA product of the mutant gene, is the 'genomic blot'.[56] This technique allows alterations in a structural gene to be detected by hybridisation of the ^{32}P-labelled plasmid DNA to nitrocellulose filters on which size-fractionated restriction fragments of normal and mutant nuclear DNA have been immobilised.[57]

Most of the techniques discussed in this chapter could, in theory, be applied to any collection of cloned DNA fragments. In practice, however, the procedures are generally only suitable for screening fairly small libraries, comprising not more than a few thousand clones. As mentioned in Chapter 9, the screening of a 'genomic' library consisting of perhaps 200,000 cloned fragments of nuclear DNA, requires the availability of a highly specific hybridisation probe, usually in the form of a cloned cDNA sequence.

Further Reading

D.M. Glover, *Genetic Engineering: Cloning DNA* (Chapman and Hall, London, 1980).

R.W. Old and S.B. Primrose, *Principles of Gene Manipulation: An Introduction to Genetic Engineering* 2nd Ed. (Blackwell Scientific Publications, Oxford, University of California Press, Berkeley and Los Angeles, 1981).

R.F. Schleif and P.C. Wensink, *Practical Methods in Molecular Biology* (Springer-Verlag, New York).

J.K. Setlow and A. Hollaender (eds.), *Genetic Engineering: Principles and Methods* (Plenum Press, New York), vols. 1 and 2.

H.O. Smith, D.B. Danner and R.A. Deich, 'Genetic transformation', *Ann. Rev. Biochem.*, 50 (1981), 41–68.

R. Williamson (ed.), *Genetic Engineering* (Academic Press, New York), Vols. 1 and 2.

References

1. M. Mandel and A. Higa, 'Calcium-dependent bacteriophage DNA infection', *J. Mol. Biol., 53* (1970), 159–62.
2. S.N. Cohen, A.C.Y. Chang and L. Hsu, 'Nonchromosomal antibiotic resistance in bacteria: genetic transformation of *Escherichia coli* by R-factor DNA', *Proc. Natl. Acad. Sci USA, 69* (1972), 2110–14.
3. E.M. Lederberg and S.N. Cohen, 'Transformation of *Salmonella typhimurium* by plasmid deoxyribunucleic acid', *J. Bacteriol., 119* (1974), 1072–74.
4. M. Dagert and S.D. Ehrlich, 'Prolonged incubation in calcium chloride improves the competence of *Escherichia coli* cells', *Gene, 6* (1979), 23–8.
5. S.R. Kushner, 'An improved method for transformation of *Escherichia coli* with ColE1 derived plasmids', in H.W. Boyer and S. Nicosia (eds.), *Genetic Engineering* (Elsevier/ North Holland Biomedical Press, Amsterdam, 1978), 17–23.
6. D.A. Morrison, 'Transformation and preservation of competent bacterial cells by freezing', *Methods in Enzymology, 68* (1979), 326–31.
7. J.F. Morrow, 'Recombinant DNA techniques', *Methods in Enzymology, 68* (1979), 3–24.
8. Ibid.
9. M. Grunstein and D.S. Hogness, 'Colony hybridization: a method for the isolation of cloned DNAs that contain a specific gene', *Proc. Natl. Acad. Sci. USA, 72* (1975), 3961–65.
10. M. Grunstein and J. Wallis, 'Colony hybridization', *Methods in Enzymology, 68* (1979), 379–89.
11. R.E. Thayer, 'An improved method for detecting foreign DNA in plasmids of *Escherichia coli*', *Anal. Biochem., 98* (1979), 60–3.
12. D. Hanahan and M. Meselson, 'High density plasmid screening', *Gene, 10* (1980), 63–7.
13. Thayer, 'An improved method for detecting foreign DNA'.
14. B. Weiss and C. Milcarek, 'Mass screening for mutants with altered DNases by microassay techniques', *Methods in Enzymology, 29* (1979), 180–96.
15. M. Noda, Y. Furutani, H. Takahashi, M. Toyosato, T. Hirose, S. Inayama, S. Nakanishi and S. Numa, 'Cloning and sequence analysis of cDNA for bovine adrenal preproenkephalin', *Nature, 295* (1982), 202–6.
16. P. Hudson, J. Haley, M. Cronk, J. Shine and H. Niall, 'Molecular cloning and characterization of cDNA sequences coding for rat relaxin', *Nature, 291* (1981), 127–31.
17. J.W. Szostak, J.I. Stiles, B.-K. Tye, P. Chiu, F. Sherman and R. Wu, 'Hybridization with synthetic oligonucleotides', *Methods in Enzymology, 68* (1979), 419–28.
18. G. Chaconas and J.H. Van de Sande, '5′-^{32}P labeling of RNA and DNA restriction fragments', *Methods in Enzymology, 65* (1980), 75–85.
19. B.G. Forde, M. Kreis, M.B. Bahramian, J.A. Matthews, B.J. Miflin, R.D. Thompson, D. Bartels and R.B. Flavell, 'Molecular cloning and analysis of cDNA sequences derived from poly A$^+$ RNA from barley endosperm: identification of B hordein related clones', *Nucleic Acid Res., 9* (1981), 6689–707.
20. Grunstein and Hogness, 'Colony hybridization'.
21. J.R. Bedbrook, S.M. Smith and R.J. Ellis, 'Molecular cloning and sequencing of cDNA encoding the precursor to the small subunit of chloroplast ribulose-1,5-bisphosphate carboxylase', *Nature, 287* (1980), 692–7.
22. Ibid.
23. G.R. Crabtree and J.A. Kant, 'Molecular cloning of cDNA for the α, β, and γ chains of rat fibrinogen: a family of coordinately regulated genes', *J. Biol. Chem., 256* (1981), 9718–23.
24. Grunstein and Wallis, 'Colony hybridization'.
25. F.C. Kafatos, C.W. Jones and A. Efstratiadis, 'Determination of nucleic acid sequence homologies and relative concentrations by a dot blot hybridization procedure', *Nucleic Acids Res., 7* (1979), 1541–52.
26. R.W. Davis, M. Thomas, J. Cameron, T.P. St John, S. Scherer and R.A. Padgett, 'Rapid DNA isolations for enzymatic and hybridization analysis', *Methods in Enzymology, 65* (1980), 404–11.
27. J. Telford, P. Bosele, W. Schaffner and M. Birnsteil, 'Novel screening procedure for

recombinant plasmids', *Science, 195* (1977), 391–2.
28. W.M. Barnes, 'Plasmid detection and sizing in single colony lysates', *Science, 195* (1977), 393–4.
29. E.M. Southern, 'Detection of specific sequences among DNA fragments separated by gel electrophoresis', *J. Mol. Biol., 98* (1975), 503–17.
30. B.M. Paterson, B.E. Roberts and E.L. Kuff, 'Structural gene identification and mapping by DNA.mRNA hybrid-arrested cell-free translation', *Proc. Natl. Acad. Sci. USA, 74* (1977), 4370–74.
31. P. Beard, J.F. Morrow and P. Berg, 'Cleavage of circular superhelical simian virus 40 DNA to a linear duplex by S1 nuclease', *J. Virol., 12*, 1303–13.
32. M. Cochet, F. Perrin, F. Gannon, P. Chambon, G.S. McKnight, D.C. Lee, K.E. Mayo and R. Palmiter, 'Cloning of an almost full-length chicken conalbumin double-stranded cDNA', *Nucleic Acids Res., 6* (1979), 2435–52.
33. R.P. Ricciardi, J.S. Miller and B.E. Roberts, 'Purification and mapping of specific mRNAs by hybridization-selection and cell-free translation', *Proc. Natl. Acad. Sci. USA, 76* (1979), 4927–31.
34. D.F. Smith, P.F. Searle and J.G. Williams, 'Characterisation of bacterial clones containing DNA sequences derived from *Xenopus laevis* vitellogenin mRNA', *Nucleic Acids Res., 6* (1979)–, 487–506.
35. Forde, Kreis, Bahramian, Matthews, Miflin, Thompson, Bartels and Flavell, 'Molecular cloning and analysis of cDNA'.
36. Ibid.
37. Smith, Searle and Williams, 'Characterisation of bacterial clones'.
38. Cochet, Perrin, Gannon, Chambon, McKnight, Lee, Mayo and Palmiter, 'Cloning of double stranded cDNA'.
39. R. Derynck, J. Content, E. DeClercq, G. Volckaert, J. Tavernier, R. Devos and W. Fiers, 'Isolation and structure of a human fibroblast interferon gene', *Nature, 285* (1980), 542–7.
40. D.W. Cleveland, S.G. Fischer, M.W. Kirschner and U.K. Laemmli, 'Peptide mapping by limited proteolysis in sodium dodecyl sulfate and analysis by gel electrophoresis', *J. Biol. Chem., 252* (1977), 1102–6.
41. J.C. Alwine, D.J. Kemp, B.A. Parker, J. Reiser, J. Renart, G.R. Stark and G.M. Wahl, 'Detection of specific RNAs or specific fragments of DNA by fractionation in gels and transfer to diazobenzyloxymethyl paper', *Methods in Enzymology, 68* (1980), 220–42.
42. P.S. Thomas, 'Hybridization of denatured RNA and small DNA fragments transferred to nitrocellulose', *Proc. Natl. Acad. USA, 77* (1980), 5201–05.
43. P.W.J. Rigby, M. Dieckmann, C. Rhodes and P. Berg, 'Labeling deoxyribonucleic acid to high specific activity *in vitro* by nick translation with DNA polymerase 1', *J. Mol. Biol., 113* (1977), 237–51.
44. A.R. MacLeod, J. Marn and S. Brenner, 'Molecular analysis of the *unc*-54 myosin heavy-chain gene of *Caenorhabditis elegans*', *Nature, 291* (1981), 386–90.
45. D. Anderson, L. Shapiro and A.M. Skalka, 'In situ immunoassays for translation products', *Methods in Enzymology, 68* (1979), 428–36.
46. L. Clarke, R. Hitzeman and J. Carlton, 'Selection of specific clones from colony banks by screening with radioactive antibody', *Methods in Enzymology, 68* (1979), 436–42.
47. H.A. Erlich, S.N. Cohen and H.O. MacDevitt, 'Immunological detection and characterization of products translated from cloned DNA fragments', *Methods in Enzymology, 68* (1979), 443–53.
48. Morrow, 'Recombinant DNA techniques'.
49. Anderson, Shapiro and Skalka, 'In situ immunoassays'.
50. Erlich, Cohen and MacDevitt, 'Immunological detection'.
51. J.C. Fiddes and H.M. Goodman, 'The cDNA for the β-subunit of human chorionic gonadotropin suggests evolution of a gene by readthrough into the 3'-untranslated region', *Nature, 286* (1980), 684–7.
52. Davis, Thomas, Cameron, St John, Scherer and Padgett, 'Rapid DNA isolations'.
53. Fiddes and Goodman, 'The cDNA for the β-subunit'.
54. H. Garoff, H. Riedel and H. Lehrach, 'A procedure to verify an amino acid sequence which has been derived from a nucleotide sequence: application to the 26S RNA of Semliki Forest virus', *Nucleic Acids Res., 10*(1982), 675–87.
55. M.G. Persico, D. Toniolo, C. Nobile, M. D'Urso and L. Luzzatto, 'cDNA sequences of

human glucose 6-phosphate dehydrogenase cloned in pBR322', *Nature, 294* (1981), 778–80.
56. A.J. Jeffreys and R.A. Flavell, 'A physical map of the DNA regions flanking the rabbit β-globin gene', *Cell, 12* (1977), 429–39.
57. B. Burr and F.A. Burr, 'Controlling element events at the shrunken locus in maize', *Genetics, 98* (1981), 143–56.

13 THE USE OF MINICELLS AND MAXICELLS TO DETECT THE EXPRESSION OF CLONED GENES

Bauke Oudega and Frits R. Mooi

CONTENTS

1. General Introduction

An analysis of cloned DNA fragments involves the characterisation of the products of gene expression and the regulation of their synthesis. When it is known that a DNA fragment codes for a particular protein or RNA species, suitable tests can be devised to determine if the cloned DNA is expressed in a particular host and the subcellular localisation of the products can be investigated. Thus, proteins can be detected by sensitive immunoassays, RNA by hybridisation to suitable probes and enzymes by appropriate enzyme assays. But how does one detect expression of genes on cloned DNA fragments when the function of the gene products is not known or if no suitable detection assays are available? The answer is to look for the synthesis of novel polypeptides or RNA species in cells carrying the cloned DNA fragments. However in wild-type cells the detection of new proteins and RNA species is almost impossible because of the concurrent expression of the host genome. Three approaches have been devised to circumvent this problem and to detect expression of cloned DNA (plasmid DNA) borne gene products: DNA-driven *in vitro* protein and RNA synthesis; minicells and maxicells.

In vitro synthesis of protein is described in Chapter 7. The use of minicells and maxicells for the identification and subcellular localisation of cloned DNA encoded gene products will be described in this chapter.

2. The Use of Minicells

Introduction

Minicells are small non-growing bodies produced by aberrant cell division at the polar ends of bacteria (Figure 13.1). They are anucleate or chromosome-DNA deficient and approximately spherical in shape. Minicells are produced either rarely in wild-type strains of bacteria or continuously during growth of mutant strains of bacteria. The first report of a minicell-producing strain was made by Gardner[1] in 1930, who described an unequal cell division process in a strain of *Vibrio cholera*. More recently Adler *et al.*[23] characterised a minicell-producing mutant strain of *E. coli* in some detail, introduced the term 'minicell', and underlined the potential usefulness of minicells and minicell-producing mutants. Subsequently, many minicell-producing strains have been isolated in gram-negative and gram-positive bacteria including *E. coli*, *Salmonella* species, *Bacillus subtilis*, and *Haemophilus influenza*.[4] Although minicells do not have a chromosome, they usually contain several copies of the plasmid(s) present in the parent cells. However, not all plasmids segregate into minicells to the same extent and some plasmids are unable to segregate into minicells at all. For instance, a 130×10^6 dalton cryptic plasmid in a *S. typhimurium* minicell producer seems to be unable to segregate into minicells.[5]

Figure 13.1: Electron Micrograph of Minicell-producing Strain E. coli K-12 DS 410 *grown to the Stationary Phase in Brain Heart Infusion Medium*

Note: The arrows indicate the position of minicells.

Plasmid-containing minicells are able to synthesise DNA, RNA, protein and cell envelope components.[6,7] Plasmids can be introduced in a minicell-producing strain either by conjugation or transformation (see Chapter 12).

Alternatively, but probably more laborious, a minicell-producing mutant strain can be isolated from a wild-type strain already harbouring the plasmid or cloned DNA to be studied.[8] Minicells produced by these mutant strains will 'automatically' contain the plasmid or cloned DNA. Other methods of introducing a DNA molecule into minicells are direct conjugation between a recipient minicell and a 'normal' doner cell,[9] or infection of minicells with bacteriophages.[10]

Introduction of a plasmid, cloned DNA, or a phage genome results in the synthesis of RNA and protein, and permits their analysis in the absence of labelled host cell products. Therefore, because minicells do not contain chromosomal DNA but only plasmid DNA, they are very suitable for analysis of newly synthesised plasmid encoded gene products. More information about the production, properties and utility of bacterial minicells can be found in a review by Frazer and Curtiss III.[11]

Purification of Minicells

Table 13.1 contains a list of minicell-producing strains of *E. coli, S. typhimurium* and *Bacillus subtilis*. This list is not complete but it provides information on some of the more important minicell-producing strains available. These strains are constructed by conventional methods for the isolation of specific mutant derivatives.[12] When no minicell-producing strains are available, a minicell-producing mutant strain can be isolated as described elsewhere.[13] The expression of the minicell-producing phenotype in newly constructed or 'old' strains is somewhat variable in both *E. coli* and *S. typhimurium*. Minicell-producing clones may vary with respect to the uniformity of size of minicells produced, the number of minicells produced, and the size and length of the parental cells. Therefore it is important to select a good clone by phase-contrast microscopy and sucrose gradient centrifugation.[14] In the course of a study it may be periodically necessary to re-isolate a 'good' minicell producer as described in Frazer and Curtis.[15] Once a 'good' minicell-producing clone has been isolated, portions of an overnight culture of that strain can be stored at −20°C in the presence of 30% glycerol.

Minicells are produced throughout the growth cycle, at all temperatures at which growth occurs and in all types of liquid and solid media by the minicell-producing strains shown in Table 13.1.[16] Minicell production is most frequent in the late-log to early-stationary phase of growth. The yield and purity of minicells produced also varies with the growth medium such that higher yields are observed on media that allow faster growth rates.[20]

Many techniques, such as differential filtration, differential centrifugation, have been used to purify minicells.[21] When highly purified preparations are desired, a combination of purification procedures can be used. It is critical to determine the level of parental cell contamination in purified minicell suspensions by viable counts on, for instance, Brain Heart Infusion agar plates or by phase-contrast microscopy. An effective method

Table 13.1: Minicell-producing Strains of E. coli, S. typhimurium *and* B. subtilis

Genus and species	Strain designation	Genotype
E. coli K–12	DS410[a]	prototroph, λ^s
	P678–54[b]	*thr, leu, lacY, minA*, $T6^s$, *gal, minB*, str^r, *thi.*
	λ925[c]	F^-, *thr, ara, leu,* azi^r, *lonA, lacY*, $T6^s$, *minA, galλ^-, minB*, str^r, *malA, xyl, mtl, thi, sup.*
	χ974[c]	F^-, *thr, ara, leu*, $T6^s$, *minA*, λ^-, *minB, str, malA, xyl, mtl, dnaB*, (TS)*sup.*
	χ964[c]	F^+, $T6^s$, *minA*, λ^-, *minB*, str^s.
	χ984[c]	F^-, $T6^s$, *minA, purE*, λ^-, *pdxC, minB, his*, str^r, $T3^r$, *xyl, ilv*, $cycA^r$.
	χ1081[c]	F^-, *thr, lacY, proC*, $T6^r$, *minA, pvrE*, λ^-, *minB, his*, str^r, $T3^r$, *xyl, ilv*, $cycA^r$, $cycB^r$, *met.*
	χ1298[c]	F^-, *dapD*, $T6^s$, *minA, purE*, λ^-, *pdxC, minB, his*, str^r, $T3^r$, *xyl, ilv, cycA, cycB, met.*
	P4121[c]	F^-, *lac, minA, minB, recA, argG, mal, xyl, metB, thi.*
S. typhimurium	χ1313[c]	prototroph, lysogenic for 3 phages
B. subtilis	VA51[c]	*thr, trpC, divA51.*
	VA356[c]	*thr, trpC, divA35.*
	VA27[c]	*thr, trpC-, divA27.*
	GSY1037[c]	*ura, recA, divIVA1.*
	RUB770[c]	*leuA, divIVA1.*
	BR59–*divIVA1*[c]	*trpC2, pheA2, divIVA1.*
	BR59–*divIVB1*[c]	*trpC2, ilvC, divIVB1.*
	M11[c]	*uru, met, divIVB1.*
	CU403–*divIVA1*[c]	*thyA, thyB, ilvC, divIVA1.*
	CU403–*divIVB1*[c]	*thyA, thyB, met, divIVB1.*
	CU403–*divIVA1*[c] –*tag-1*	*thyA, thyB, ilvC, divIVA1- tag-1.*
	CU403–*divIVB1*[c] *tag-1*	*thyA, thyB, metB, div IVB1, tag-1.*

Notes: a. See Dougan and Sheratt.[17]
b. See Adler *et al.*[18]
c. See Frazer and Curtiss III[19] for all the following examples.

for the purification of minicells is the use of three successive linear 5 to 30% w/v sucrose in buffered saline gradients. Cells and minicells grown to the late-log or stationary phase (an overnight culture) are collected by centrifugation (10 minutes, $15{,}000 \times g$) and the cells and minicells are resuspended in buffered saline such that 50- to 60-fold concentrated suspensions are obtained. These concentrated suspensions should always be vortexed vigorously for a couple of minutes or resuspended by several passages through a syringe or pipette prior to layering on top of a sucrose gradient to eliminate aggregates and to thus reduce the loss of minicells by co-sedimentation with cells and to increase the yield of the minicell suspension. The suspensions (4 ml) are layered on top of 80 ml of a 5 to 30% w/v sucrose gradients in clear, polycarbonate centrifuge bottles. These gradients can easily be prepared by freezing a centrifuge bottle containing 80 ml of 20% w/v sucrose in buffered saline. After slowly thawing (for instance overnight at 4°C) 5–30% gradients have been formed. The bottles are centrifuged at 0°C at 3000 × g in a swinging-bucket rotor in a bench-top centrifuge. After centrifugation (30 minutes), the minicell bands at approximately one-third of the distance from the top of the gradients are conveniently withdrawn from the gradients by means of a syringe. The minicell suspension is then slowly diluted in buffered saline at 0°C. The minicells are pelleted by centrifugation (10 minutes; 35,000 × g rev/min) and resuspended in a small volume of buffered saline. The sucrose density centrifugation step should be repeated twice on 40 ml of a 5 to 30% gradient and finally the minicells are suspended in label medium to an optical density at 660 nm of 0.4 (about 3×10^9 minicells/ml). Excellent purification results are obtained if, rather than the complete minicell bands, only the upper two thirds of the bands are withdrawn from the gradients. Minicell suspensions prepared by this method normally contain less than one viable cell per 10^6–10^7 minicells. Other methods of purification, such as differential filtration, differential centrifugation, sonication or u.v.-irradiation of viable cells present in a minicell suspension, can also be employed but are usually more laborious. For labelling, minicells are resuspended in label medium to an optical density of about 0.4 at 660 nm. Label medium is usually composed of M9 medium plus supplements such as glucose (0.4%), vitamins or specific amino acids. In case where minicells are labelled with ^{35}S-methionine, ^{35}S-methionine assay medium (Difco) can be added to the label medium. Minicells can also be suspended in buffered saline plus 30% glycerol and stored at –20, –70, or –190°C. Storage, however, will result in a decreased DNA, RNA and protein synthesising capacity[22] (especially storage at – 20°C).

Radioactive Labelling of Polypeptides

Having introduced the recombinant DNA under analysis into a minicell-producing strain (see introduction to Section 2) it is then necessary to detect the products of gene expression. Minicells isolated from a strain harbouring the recombinant DNA will contain that DNA. Plasmid-containing minicells are able to incorporate radioactive precursors for RNA (uridine and uracil) and protein (amino acid mixtures, specific amino acids) synthesis into TCA-insoluble material.[23] In general the rate of incorporation of precursors into protein by minicells is only one to two percent of the rate observed with a cell suspension of the same optical density. Different plasmid-containing minicells have different capacities for the incorporation of precursors.[24]

RNA and protein synthesis can be inhibited by specific inhibitors such as rifampin (inhibitor of RNA synthesis) and chloramphenicol (inhibitor of protein synthesis).

Minicells that harbour no plasmid or cloned DNA exhibit a small amount of incorporation of precursors into protein and into RNA. Proteins synthesised in these minicells have been seen as small radioactive bands or as random high background on sodium dodecyl sulphate (SDS) polyacrylamide gels.[25] The small amount of protein synthesised by minicells purified at room temperature (20–23°C) is greatly reduced if minicells are purified at 0°C.[26] The low level of protein synthesis in DNA-deficient minicells may be due to persistence of stable cellular messenger RNA. Alternatively, this synthesis may be due to transcription of small amounts of random cellular DNA trapped in the minicells. Minicell preparations with a large number of contaminating parental cells show a higher background of proteins on SDS polyacrylamide gels.

For labelling, minicells are suspended in label medium (M9 medium plus supplements[27]) and pre-incubated for a short period of time (15 minutes) at the desired temperature (e.g. 37°C) in order to degrade chromosome encoded messenger RNA possibly present in the minicells. Then the radioactive labelled precursor is added (10–100 μCi/ml) and the incubation is continued. In our laboratory, minicells are routinely labelled for two hours, but shorter label-periods (10–20 minutes) or overnight incubation are also possible. Label temperature may vary between 20° and 40°C.

It must be emphasised that minicells differ from normal cells in that not all induction or derepression systems function in minicells as in normal cells. Not all plasmid or recombinant DNA is expressed in minicells.[28] Obviously, the possible interaction between plasmid encoded components, and chromosome encoded components active in promoting or repressing the plasmid functions, may be quite complex and absent in minicells. Thus, much variability among plasmids with respect to expression of specific plasmid genes in minicells has to be expected.[29,30]

Analysis of Labelled Polypeptides

SDS polyacrylamide gel electrophoresis is carried out to obtain profiles of polypeptides synthesised in minicells. After labelling minicells or minicell-extracts are precipitated with 5–10% trichloroacetic acid and the insoluble pellet is washed with ethanol or acetone and solubilised in a small volume of sample buffer for SDS polyacrylamide gel electrophoresis by heat-treatment.[31,32] Minicells can also be pelleted by centrifugation and solubilised in sample buffer by a heat-treatment (5 minutes, 100°C). SDS polyacrylamide gel electrophoresis is usually carried out according to the procedures described by Laemmli[33] or Lugtenberg[34] (see also Chapter 2). After electrophoresis the gels are fixed, stained (optional) and dried. The plasmid DNA encoded polypeptide bands can be visualised by fluorography or by autoradiography. Fluorography is necessary when minicells have been

Figure 13.2: Analysis of polypeptides and precursors of polypeptides encoded by pFM205 and its deletion derivatives pFM55 and pFM201

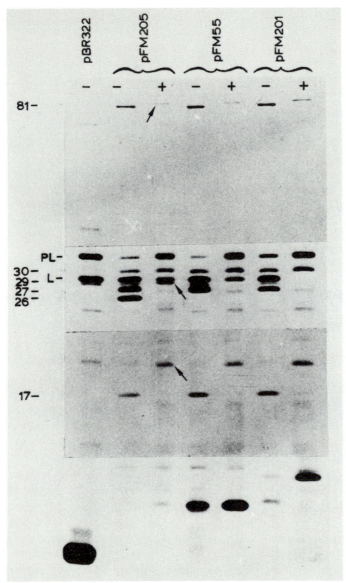

Note: Plasmid pFM205 is composed of a 4.3-megadalton large DNA fragment derived from the wild-type K88ab plasmid pRI8801. Plasmids pFM55 and pFM201, deletion derivatives of pFM205, were constructed as described by Mooi *et al.*[35] Minicells were labelled with ^{35}S-methionine in the absence (−) or presence (+) of 9.5% ethanol and analysed on a 24-cm-long 14 to 20% SDS polyacrylamide gradient gel.[36] The location of the pBR322-encoded β-lactamase (L) and its precursor (PL) and the molecular masses in Kd of the polypeptides expressed by the cloned DNA are indicated at the left. The arrows in the third lane indicate the precursors of the 81, 26, and 17 Kd polypeptides.

labelled with tritiated amino acids and is optional if ^{14}C- or ^{35}S-amino acids are used. Small polypeptides or polypeptides which are difficult to separate can be analysed on linear or exponential SDS polyacrylamide gradient gels. Figure 13.2 shows an analysis of polypeptides synthesised in *E. coli* minicells separated on a SDS polyacrylamide gradient gel.

In minicells, processing of polypeptides synthesised with a N-terminal signal or leader sequence[37] is not fully effective (see Figure 13.2). Therefore one gene can code for two polypeptides, the precursor and the processed polypeptide. As in normal cells, processing in minicells can be inhibited by membrane perturbants such as ethanol.[38] This is an elegant method to detect polypeptides that are synthesised as a precursor with a N-terminal signal peptide (Figure 13.2).

The analysis of labelled polypeptides on SDS polyacrylamide gels allows for the establishment of the apparent molecular weight of the polypeptides under denaturing conditions. In case plasmid or cloned DNA encoded polypeptides synthesised in minicells have to be studied in their biological active configuration, gel electrophoresis of minicell extracts can also be carried out under non-denaturing conditions.

3. The Use of Maxicells

Introduction

When heavily irradiated *E. coli* cells are infected with Lambda phages, or its transducing derivatives the radioactive protein label added to the medium after infection is preferentially incorporated into phage coded polypeptides because of the inability of the damaged bacterial DNA to act as a template for transcription.[40,41] This method is very useful in the identification of products of *E. coli* genes incorporated in transducing phages, but it cannot be applied to plasmids or cloned DNA (recombinant DNA) because the frequency of transformation of *E. coli* cells with plasmids is at most 10^{-2}–10^{-3}.[42] A relatively simple, alternative method for preferentially labelling and identifying plasmid and cloned DNA encoded gene products, is the use of maxicells. The maxicell system is based on two observations. First, when irradiated with u.v. light (254 nm), *E. coli recA, uvrA* cells stop DNA synthesis and chromosomal DNA is extensively degraded so that only a small amount of the chromosomal DNA remains several hours after irradiation.[43] Secondly, if these cells contain a multicopy plasmid, the plasmid molecules that did not receive a u.v. hit continue to replicate with plasmid DNA levels increasing about 10-fold within six hours after irradiation in cells where over 80% of the chromosomal DNA was degraded.[44] Hence these cells , maxicells, contain mostly plasmid DNA and synthesise almost exclusively plasmid proteins. The success of the maxicell system is critically dependent upon the different target size of plasmids and the chromosome. The maximum size of plasmids that can be studied in

Figure 13.3: Analysis of polypeptides encoded by pFM205 (see Figure 13.2) in the various subcellular fractions

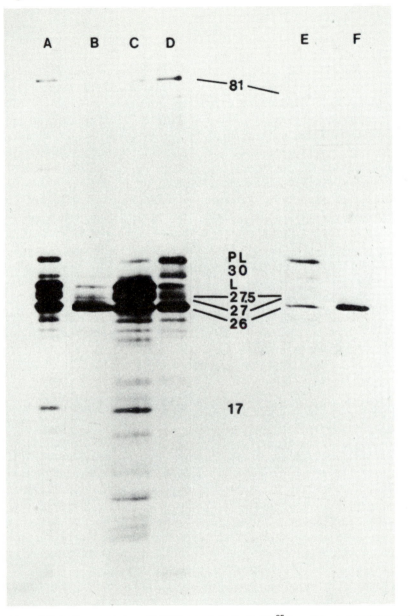

Note: Minicells containing pFM205 were labelled with ^{35}S-methionine and the various subcellular fractions were prepared as described by Van Doorn et al.[39] The apparent molecular masses in Kd of the polypeptides encoded by pFM205 and the positions of β-lactamase (L) and its precursor (PL) are indicated. (A) Unfractionated minicells; (B) cytoplasmic fraction; (C) periplasmic fraction; (D) total membrane fraction; (E) purified outer membrane fraction; (F) purified inner membrane fraction.

maxicells is not known exactly, but plasmids with a molecular weight of up to 10×10^6 can be used.[45]

Radioactive Labelling of Maxicells

E. coli CSR 603 (*rec A1, uvr A6, phr-1*) is used for maxicell experiments.[46] This strain has a Phr (non-photoreactivable) phenotype which allows the experiments to be carried out under ordinary laboratory lighting.[47] Plasmid or cloned DNA can be introduced by conjugation or transformation. This strain and its plasmid-containing derivatives are routinely grown in M9 medium supplemented with 1% Casamino acids to 2×10^8 cells per ml and irradiated with a u.v. fluence of about 50 Jm^{-2} from an u.v. lamp at a fluence rate of 0.5 $Jm^{-2}s^{-1}$. The irradiated cells are then incubated at 37°C with shaking for 16 hours. Then samples of the culture are centrifugated, the cells are washed with M9 medium and resuspended in minimal medium. After a starvation period of about one hour radioactive amino acids (^{3}H- or ^{14}C-labelled amino acid mixtures or specific labelled amino acids such as ^{35}S-methionine) are added and incubation is continued for another hour. The labelled cells are then collected by centrifugation and solubilised in sample buffer for SDS polyacrylamide gel electrophoresis.[48,49] Analysis of labelled products can be conducted as described for minicells.

It must be emphasised that not all genes are expressed in maxicells. Genes that are expressed at very low levels in normal cells, due to the presence of specific repressors, ineffective promotors, or poorly translated messenger RNAs, are probably also expressed at very low levels in maxicells.

4. Subcellular Localisation of Polypeptides in Minicells and Maxicells

Introduction

The subcellular localisation of polypeptides and proteins can provide information on their biological function. In general, the compartmentation of proteins and polypeptides is studied by isolation of the various subcellular fractions: cytoplasm, inner membranes periplasm, and outer membranes. The periplasm is defined as the fluid between the inner and outer membranes. The space between the two membranes is the periplasmic space containing the periplasm. In these isolated fractions the presence of enzyme activities or labelled polypeptides can be examined.

Various investigations on the localisation of chromosomal DNA encoded enzymes in minicells showed that the compartmentation of these enzymes is essentially identical to that in cells.[50] However the specific activity of certain periplasmic enzymes is higher in *E. coli* minicells than in cells, whereas the specific activity of cytoplasmic enzymes is lower.[51] An explanation for these observations is that the surface to volume ratio is higher for minicells than for cells, and that the minicells are formed at the polar ends of the cells. These polar ends might be enriched in certain periplasmic enzymes.[52]

Minicells and maxicells can be used to study the subcellular compartmentation of plasmid or cloned DNA encoded proteins and polypeptides.[53,54] In general, all techniques used to isolate the various subcellular fractions of cells can be applied to minicells or maxicells. However, minicells do not behave as normal cells in particular experiments probably as a result of their higher surface to volume ratio and/or their origin of formation. Most of the localisation studies conducted so far concerned minicells of *E. coli*. In the following chapters the various techniques used to compartmentalise minicells of *E. coli* are discussed. It can be expected that maxicells behave in localisation studies as stationary *E. coli* cells.

Separation of Membrances and Soluble Protein Fraction

Cells and minicells of *E. coli* can be broken by sonic disruption, rupture by a French pressure cell or rupture following treatment with lysozyme in the absence or presence of a detergent. Minicells are more resistant to sonic rupture than normal cells. Satisfactory results are obtained by treatment of the minicells for longer time periods, avoiding overheating the minicell suspension.[55] Minicells in phosphate or Tris-HCl buffer (ph 7–8) are sonicated at high power for 30-sec pulses in the cold for 5–6 min. Normally cells are sonicated for two to three minutes. Following sonication the optical density of the minicell suspension or microscopy can provide information on the percentage of unbroken minicells. Usually 80–90% breakage is obtained by sonication of minicells. Unbroken minicells are removed by low-speed centrifugation of the broken minicell suspension (5 min, $10,000 \times g$) and subsequently the broken minicell suspension (a mixture of membranes and soluble proteins) is centrifuged at high speeds (2 h, $190,000–200,000 \times g$) to collect the total membrane fraction (inner and outer membranes). The supernatant fluid contains the soluble protein fraction. The broken minicell suspension, the total membrane fraction, and the soluble protein fraction (periplasm and cytoplasm) can be used for enzyme assays or analysed on denaturing or non-denaturing polyacrylamide gels.

Another adequate method of lysing minicells is rupture by passages through a French pressure cell.[56] Minicells must be passed through such a pressure cell under 16,000 to 20,000 psi in order to obtain maximum breakage, while 12,000–14,000 psi is adequate for cells. Other methods such as grinding with quartz glass or treatment of the minicells with lysozyme with or without EDTA and detergents have also been employed.[57] However, a satisfactory separation of membrane-bound and soluble protein is not possible after treatment of minicells with detergents.

Isolation of Periplasmic and Cytoplasmic Fractions

The soluble protein fraction of gram-negative cells and minicells consists of a periplasmic and a cytoplasmic fraction. In order to localise enzymes or polypeptides in these fractions two methods have been used.[58] First, periplasmic proteins are released during the formation of spheroplasts and secondly periplasmic proteins can be released by an osmotic shock procedure. A satisfactory method to convert cells of *E. coli* to spheroplasts has been published by Witholt *et. al.*[59] In this method cells are treated with lysozyme in the presence of EDTA and a small osmotic shock is used to transfer the lysozyme molecules across the outer membrane into the periplasmic space where the peptidoglycan is degraded. This method is also suitable for minicells provided that some minor changes are made in the protocol. Treatment of minicells with increased concentrations of EDTA and egg white lysozyme for 30 minutes in the cold results in the formation of mini-spheroplasts. The completion of spheroplast formation, usually 80–90%, and the stability of the mini-spheroplasts can be checked by their osmotic sensitivity as described by Witholt *et al.*[60] Following spheroplast formation, $MgCl_2$ is added to stabilise the spheroplasts and the contaminating cells and spheroplasts are spun down (10 min, 5000 rev/min). The supernatant fluid contains the periplasmic proteins. The cytoplasmic fraction can be obtained by resuspending the spheroplasts and subsequent lysis of the spheroplasts by sonic oscillation, a passage through a French pressure cell, or by an osmotic shock. The membranes are then removed by centrifugation (2 h, $150,000–200,000 \times g$) and the supernatant fluid contains the cytoplasmic proteins. These fractions can be used for enzyme assays or labelled proteins and polypeptides can be analysed on denaturing or non-denaturing polyacrylamide gels following dialysis and concentration of the fractions. As controls for an adequate separation of periplasmic and cytoplasmic proteins, particular marker enzymes can be assayed. β-Galactosidase can be used as a cytoplasmic marker enzyme,[61] whereas β-lactamase[62] or alkaline phosphatase[63] can be used as periplasmic marker enzymes.

Another well-studied method of preparing spheroplasts, especially from minicells or cells of *S. typhimurium*, is published by Osborn *et al.*[64]

Periplasmic proteins and polypeptides can also be isolated by the osmotic shock procedure of Nossal and Heppel.[65] In this method cells are briefly exposed to sucrose containing EDTA, after which cells are sedimented and rapidly dispersed in cold water. Periplasmic enzymes and polypeptides are specifically released during this procedure. Growing cells are more susceptible to this method than stationary cells. This method has been applied to minicells of *E. coli*;[66] but the yield was low. For qualitative studies this method can be satisfactory.

Separation of Inner and Outer Membranes

In order to localise enzymes and polypeptides in the inner or outer membrane of labelled minicells or maxicells, two methods can be used. First, the inner and outer membranes can be separated and purified and secondly, the inner membranes can be solubilised with the use of detergents.

For the separation of inner and outer membranes a total membrane fraction is isolated, either by direct breakage of minicells or cells by sonic disruption or a French pressure cell, or following spheroplast formation.[67] The total membrane fraction (inner and outer membranes) is then subjected to isopycnic sucrose density gradient centrifugation according to Osborn *et al.*[68] For minicells, breakage by a French pressure cell has been successful.[69] Following separation the fractions containing the inner or outer membranes are pooled. The membranes are collected by centrifugation (2 h, 150,000–200,000 × g) and the presence of enzymes and labelled polypeptides can be examined. As controls, the amount of lipopolysaccharide and KDO (2-keto-3-deoxyoctonate) or the cytochrome content can be estimated as marker components for the outer and inner membrane, respectively.[70]

Another method to localise enzymes or labelled polypeptides in the membrane fraction is to solubilise the inner membranes with specific detergents. Triton X-100, a non-ionic detergent, has been shown to only solubilise inner membranes when Mg^{2+} ions are present.[71] Sarkosyl (sodium-lauryl sarcosinate), an ionic detergent, only solubilises inner membranes and in the presence of Mg^{2+} ions, a partial protection of the inner membranes from dissolution has been observed.[72] Treatment of a total membrane fraction with these detergents results in the separation of inner and outer membrane associated polypeptides. It should be emphasised that although both these detergents were found to solubilise inner membranes, while outer membranes were not affected, differences in the proteins and polypeptides solubilised were observed.[73] Following solubilisation the detergent soluble fraction (inner membranes) and the detergent insoluble fraction (outer membranes) can be examined for the presence of particular enzymes or labelled polypeptides. The detergents may interfere with SDS polyacrylamide gel electrophoresis and should be removed by dialysis, extraction, acetone precipitation or chromatography, prior to electrophoresis.

In Figure 13.3 an example of the subcellular localisation of plasmid encoded polypeptide in minicells of *E. coli* is presented. Another example of a compartmentation study using minicells is shown in Palva *et al.*[74]

References

1. A.D. Gardner, 'Cell-division, colony-formation, and spore formation', in P. Fildes and J.C.G. Ledingham (eds), *A system of bacteriology in relation to medicine* (HMSO, London, 1930), vol. 1, 159–76
2. H.I. Adler, W.D. Fisher and G.E. Stapleton, 'Genetic control of cell division in bacteria', *Science*, 154 (1966), 417.
3. H.I. Adler, W.D. Fisher and A.A. Hardigree, 'Miniature *Escherichia coli* cells deficient in DNA', *Proc. Nat. Acad. Sci. (USA)*, 57 (1967), 321–6.
4. A.C. Frazer and R. Curtiss III, 'Production, properties and utility of bacterial minicells',

Curr. Top. Microbiol. Immunol., 69 (1975), 1–84.

5. R.J. Sheeny, D.P. Allison and R. Curtiss III, 'Cryptic plasmids, in a minicell-producing strain of *Salmonella typhimurium*, *J. Bacteriol.*, *114* (1973), 439–44.
6. Frazer and Curtiss III, 'Production, properties and utility'.
7. G. Mertens and J.N. Reeve, 'Synthesis of cell envelope components by anucleate cells (minicells) of *Bacillus subtilis*', *J. Bacteriol.*, *129* (1977), 1198–207.
8. Frazer and Curtiss III, 'Production, properties and utility'.
9. Ibid.
10. J.N. Reeve, 'Bacteriophage infection of minicells. A general method for identification of in vivo bacteriophage directed polypeptide biosynthesis', *Molec. Gen. Genet.*, *158* (1977), 73–9.
11. Frazer and Curtiss III, 'Production, properties and utility'.
12. J. Miller, *Experiments in molecular genetics* (Cold Spring Harbor Laboratory, Cold Spring Harbor, New York, 1972), 431.
13. Frazer and Curtiss III, 'Production, properties and utility'.
14. Ibid.
15. Ibid.
16. Ibid.
17. G. Dougan and D. Sheratt, 'The transposon Tnl as a probe for studying Col El structure and function', *Mol. Gen. Genet.*, *151* (1977), 151–60
18. Adler, Fisher, Cohen and Hardigree, 'Miniature *E. coli* cells'.
19. Ibid.
20. Ibid.
21. Ibid.
22. Ibid.
23. Ibid.
24. Ibid.
25. Ibid.
26. Ibid.
27. Miller, *Experiments in molecular genetics*.
28. R.B. Meager, R.C. Tait, M. Betlach and H.W. Boyer, 'Protein expression in *E. coli* minicells by recombinant plasmids', *Cell*, *10* (1977), 521–36.
29. Frazer and Curtiss III, 'Production, properties and utility'.
30. Meager, Tait, Betlach and Boyer, 'Protein expression in *E. coli*'.
31. W.K. Laemmli, 'Cleavage of structural proteins during the assembly of the head of bacterophage of bacterophage T4', *Nature*, *277* (1970), 680–5.
32. B. Lugtenberg, J. Meijers, R. Peters, P. Van der Hoek and L. Van Alphen, 'Electrophoretic resolution of the major outer membrane protein of *Escherichia coli* into four bands', *FEBS Letts*, *58* (1975), 254–8.
33. Laemmli, 'Cleavage of structural proteins'.
34. Lugtenberg, Meijers, Peters, Van der Hoek and Van Alphen, 'Electrophoretic resolution'.
35. F.R. Mooi, F.K. De Graaf and J.D.A. Van Embden, 'Cloning, mapping and expression of the genetic determinant that encodes for the K88ab antigen', *Nucleic Acids Res.*, *6* (1979), 849–65.
36. F.R. Mooi, N. Harms, D. Bakker and F.K. De Graaf, 'Organisation and expression of genes involved in the production of the K88ab antigen', *Infection and Immunity*, *32* (1981), 1155–63.
37. D.B. Davis and P.C. Tai, 'The mechanism of protein secretion across membranes', *Nature*, *283* (1980), 433–8.
38. E.T. Palva, T.R. Hirst, S.J.S. Hardy, J. Holmgren and L. Randall, 'Synthesis of a precursor to the B subunit of heat-labile enterotoxin in *Escherichia coli*', *J. Bacteriol.*, *146* (1981), 325–30.
39. J. Van Doorn, B. Oudega, F.R. Mooi and F.K. De Graaf, 'Subcellular localisation of polypeptides involved in the biosynthesis of K88ab fimbriae', *FEMS Microbiol. Letters*, *13* (1982), 99–104.
40. M. Ptashne, 'Isolation of the λ phage repressor', *Proc. Natc. Acad. Sci. USA.*, *57* (1967), 306–13.
41. S.R. Jaskunas, L. Lindahl, M. Nomura and R.R. Burgess, 'Identification of two copies of

the gene for the elongation factor EF–Tu in *E. coli*', *Nature*, *257* (1975), 458–62
42. A. Sancar, A.M. Hack and W.D. Rupp, 'Simple method for identification of plasmid-coded proteins', *J. Bacteriol.*, *137* (1979), 692–3.
43. P. Howard-Flanders, 'Genes that control DNA repair and genetic recombination in *Escherichia coli*', *Adv. Biol. Med. Phys.*, *12* (1968), 299–317.
44. Sancer, Hack and Rupp, 'Simple method for identification of proteins'.
45. Ibid.
46. Ibid.
47. Ibid.
48. Laemmli, 'Cleavage of structural proteins'.
49. Lugtenberg, Meijers, Peters, Van der Hoek and Van Alphen, 'Electrophoretic resolution'.
50. Frazer and Curtiss III, 'Production, properties and utility'.
51. Ibid.
52. Ibid.
53. N. Kennedy, L. Beutin and M. Achtman, 'Conjugation proteins encoded by the F sex factor', *Nature*, *270* (1977), 580–5.
54. Van Doorn, Oudega, Mooi and De Graaf, 'Subcellular localisation of polypeptides'.
55. Frazer and Curtiss III, 'Production, properties and utilities'.
56. Van Doorn, Oudega, Mooi and De Graaf, 'Subcellular localisation of polypeptides'.
57. Frazer and Curtiss III, 'Production, properties and utility'.
58. Van Doorn, Oudega, Mooi and De Graaf, 'Subcellular localisation of polypeptides'.
59. B. Witholt, M. Boekhout, M. Brock, J. Kingma, H. Van Heerikhuizen and L. De Ley, 'An efficient and reproducible procedure for the formation of spheroplasts from various grown *Escherichia coli*', *Anal. Biochem.*, *74* (1976), 160–70.
60. B. Witholt, H. Van Heerikhuizen and L. De Ley, 'How does lysozyme penetrate through the bacterial outer membrane?', *Biochim. Biophys. Acta*, *443* (1976), 534–44.
61. Van Doorn, Oudega, Mooi and De Graaf, 'Subcellular localisation of polypeptides'.
62. Ibid.
63. J. Tommassen and B. Lugtenberg, 'Outer membrane protein of *Escherichia coli* K-12 is regulated with alkaline phosphatase', *J. Bacteriol.*, *143* (1980), 151–7.
64. M.J. Osborn, J.E. Gander, E. Parisi and J. Carson, 'Mechanism of assembly of the outer membrane of *Salmonella typhimurium*', *J. Biol. Chem.*, *247* (1972), 3962–72.
65. N.G. Nossal and L.A. Heppel, 'The release of enzymes by osmotic shock from *Escherichia coli* in exponential phase', *J. Biol. Chem.*, *241* (1966), 3055–62.
66. Van Doorn, Oudega, Mooi and De Graaf, 'Subcellular localisation of polypeptides'.
67. Osborn, Gander, Parisi and Carson, 'Mechanism of assembly of the outer membrane'.
68. Ibid.
69. Van Doorn, Oudega, Mooi and De Graaf, 'Subcellular localisation of polypeptides'.
70. Osborn, Gander, Parisi and Carson, 'Mechanism of assembly of the outer membrane'.
71. C.A. Schnaitman, 'Solubilisation of the cytoplasmic membrane of *Escherichia coli* by Triton X-100', *J. Bacteriol.*, *108* (1971), 545–52.
72. C. Filip, G. Fletcher, J.L. Wulff and C.F. Earhart, 'Solubilisation of the cytoplasmic membrane of *Escherichia coli* by the ionic detergent sodium-lauryl sarcosinate', *J. Bacteriol.*, *115* (1973), 717–22.
73. I. Chopra and S.W. Shales, 'Comparison of the polypeptide composition of *Escherichia coli* outer membranes prepared by two methods', *J. Bacteriol.*, *144* (1980), 425–7.
74. Palva, Hirst, Hardy, Holmgren and Randall, 'Synthesis of a precursor'.

14 GEL ELECTROPHORESIS OF DNA

Stephen A. Boffey

CONTENTS

1. Introduction

From other chapters it will be evident that the molecular biologist often needs to separate the components of a complex mixture of DNA molecules either analytically or preparatively. Although it is sometimes possible to use density or affinity techniques to separate DNA molecules on the basis of their base compositions, such an approach cannot be used with complex mixtures or when, as is often the case, different DNAs have similar base compositions. Separations based on size differences are far more generally useful. The identification of restriction fragments containing a particular sequence of bases (e.g. a particular gene) depends on the separation of DNA fragments according to size (Chapter 15), as does the mapping of the position of such a sequence in the intact DNA. Separated fragments can be recovered and then replicated indefinitely by cloning (Chapters 11, 12 and 13). The sequencing of DNA also relies on the ability to resolve nucleic acids by size (Chapter 16).

Although the digestion of DNA by certain restriction endonucleases may produce fragments only a few base pairs in length, such short pieces are not of use for cloning or sequence recognition. In general one is interested in separating molecules larger than about 200 base pairs (bp) in length, up to the size of plasmids (e.g. pBR322, 4.36 kbp; RP4, 4.55 kbp) or even whole organelle genomes (e.g. *Zea mays* chloroplast DNA, c. 135 kbp) although such large DNA molecules are extremely hard to isolate and handle without breaking by shearing.

The technique most widely used for separating DNA molecules of the sizes indicated above is electrophoresis in agarose gels. There are alternatives, but none is applicable over such a wide range of sizes; agarose gels have the advantage of being simple to prepare, and are easy to use for electrophoresis. Consequently most of this chapter will be devoted to agarose gel electrophoresis, but mention will be made of some other techniques which offer attractive alternatives for specific applications.

2. Principles of Gel Electrophoresis of Nucleic Acids

At neutral or alkaline pH the phosphate groups of DNA give rise to a uniform negative charge per unit length of the DNA molecule. Thus, in an electric field, DNA will tend to move towards the anode, with a constant driving force per unit length propelling the molecules, regardless of base composition. Any differences in the rates of movement of different DNA molecules will depend only on the resistances to their movement. If the molecules are in a gel they will have to pass through its pores as they move towards the anode. The largest molecules will have most difficulty passing through the pores, and may even be blocked completely, whereas the smallest molecules will be relatively unhindered. Consequently, the velocity of movement of a DNA molecule during gel electrophoresis will depend on its size, the smallest molecules moving fastest. This is similar to the situation in SDS polyacrylamide gel electrophoresis of SDS-denatured proteins (Chapter 2). However, the analogy is not perfect, as double-stranded (ds) DNA molecules form relatively stiff rods, and it is not completely understood how they pass through the gel. It is probable that long DNA molecules go through gel pores 'end-on'.[1] While the molecule is passing through the pore it will experience 'drag'; so the longer the molecule, the longer it will be retarded by each pore. 'Sideways' movement may become more important for very small double-stranded DNA and for the more flexible single-stranded DNA.

Because the separation depends on chain length rather than molecular weight of the molecules (being virtually independent of base composition) I will refer to DNA sizes in 'base pairs' (bp) if double-stranded, or 'nucleotides' (nt) if single-stranded. Where necessary I have converted from molecular weights by assuming 1 bp has an average molecular weight of 620 daltons. The influence of shape and size, rather than molecular weight, is demonstrated by the fact that gel electrophoresis can be used to separate different physical forms of the same DNA from each other. For example, the covalently closed circular (Form I), open circular (Form II) and linear (Form III) forms of a plasmid DNA will all have different mobilities, Form I moving fastest owing to its highly compact conformation[2] (see Figure 14.1).

3. Agarose Gels

The polyacrylamide gels used for protein fractionations (Chapter 2) have pores which are too small for the passage of large DNA molecules. Gels as dilute as 3.5% polyacrylamide are suitable for the fractionation of DNA shorter than 1000 bp, but this range cannot be extended upwards by the use of even more dilute gels unless they are strengthened by the addition of agarose. Agarose gels maintain their physical strength at much larger pore sizes than polyacrylamide gels, and so they are widely used for the fractionation of ds DNA larger than about 200 bp.

Agarose is the unbranched galactan chain component of agar (the other being agaropectin). Owing to its low sulphate content it has little charge and so results in little electro-endosmosis when used for electrophoresis. The coefficient of electro-endosmosis $(-m_r)$ is usually quoted by suppliers of agarose, and becomes smaller as the sulphate content decreases. Because the velocity of migration of a given DNA molecule is proportional to $(-m_r)^{-0.5}$ over a wide range of values of $-m_r{}^3$ it is worth using an agarose with a low $-m_r$ for normal purposes. Typical low $-m_r$ values are 0.15 or less. Agarose molecules gel by hydrogen-bonding when in cool aqueous solution, and the gel pore size is determined by the concentration of agarose; concentrations between 0.5 and 2% are commonly used, covering a DNA size range of 0.2 to 50 kbp (see below).

Figure 14.1: A Horizontal Agarose Gel viewed by u.v. Light after Electrophoresis of λDNA and the Plasmid pBR322

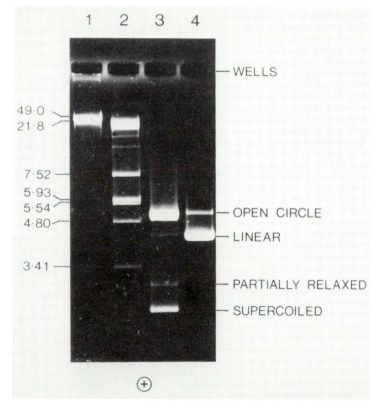

Note: DNA was visualised by staining with ethidium bromide. Track 1 – λDNA (m. wt. 49 kbp); 2 – λDNA after digestion with the endonuclease EcoRI; 3 – pBR322; 4 – pBR322 after digestion with EcoRI. Molecular weights of DNA fragments are marked in kbp.

4. Methods

Gel Preparation

The agarose is dissolved in electrophoresis buffer (Tris-acetate, Tris-phosphate or Tris-borate, pH 8, with EDTA to inhibit DNase) by heating above about 90°C; the same buffer is used both for gel preparation and in the electrode chambers. Although rod gels are sometimes used, it is most usual to cast agarose gels as vertical or horizontal slabs. Vertical gels can give slightly sharper bands than horizontal ones, but they are not quite as easy to prepare, and need a more sophisticated electrophoresis tank. Vertical gels are prepared between two plates, separated by spacers, as described for polyacrylamide gel electrophoresis (PAGE) (Chapter 2), with sample wells formed in the top of the gel using a comb. Because of the thickness of the gels (< 3 mm) and their lack of adhesion to smooth plates, they must be

Figure 14.2: A Simple Apparatus for Electrophoresis using Horizontal Agarose Gels

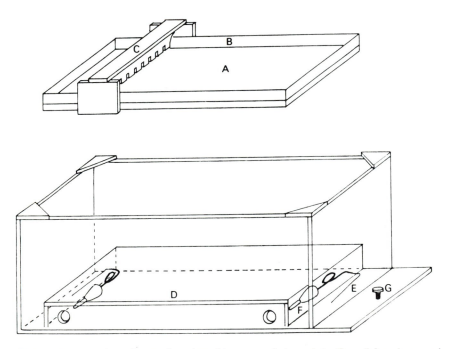

Notes: A, glass plate; B, tape forming sides around glass plate; C, well-forming comb not quite touching glass plate. The agarose solution is poured onto the glass plate with tape and comb in position. When the gel has set, the comb is removed, and the gel transferred, on its glass plate, onto support D of the electrophoresis apparatus. E, platinum electrode, connected to F, plug for connection to power supply; G, levelling screw. All parts except 'A' can be made of perspex. For safety, a lid is normally present during runs, and a microswitch can be incorporated to switch off the power supply automatically when the lid is removed.

supported by a gauze, or plates with ground glass faces must be used to stop the gel sliding out. The set gel, between its plates, is mounted in the same type of electrophoresis tank as is used for PAGE.

The preparation of horizontal gels is very simple. A temporary wall is constructed around a glass plate using adhesive tape (some use a tight-fitting plastic frame), and molten agarose is poured onto the plate. A comb is fixed in place, near one end of the plate, held about 1 mm above the glass, and is removed once the gel is set, to leave sample loading wells in the gel. This slab gel is then transferred, on its glass base, into the electrophoresis apparatus. Electrical contact between electrodes and gel can be established via filter paper wicks, or agarose bridges can be poured. However, by far the simplest method is to flood the two electrode chambers and the gel with

Figure 14.3A: Successive Stages in the Running of a Preparative Gel, viewed from above

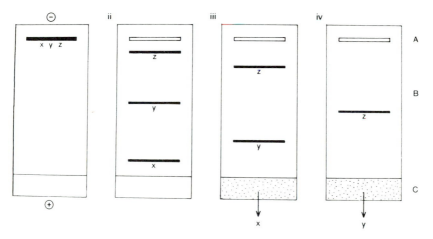

Note: (i) A mixture of nucleic acids x, y and z is loaded into the sample well A.
(ii) Electrophoresis separates the nucleic acids along the gel, B, according to size.
(iii) Further electrophoresis results in the smallest molecules (x) running off the gel into a collecting chamber, C, from which they can be recovered. (iv) Continued electrophoresis and collection from C allows the separate recovery of each nucleic acid.

Figure 14.3B: Cross-sectional Diagram of a Preparative Slab Gel Apparatus

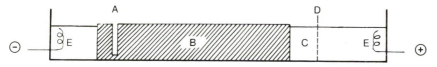

Note: A – sample loading well; B – gel (note thickness); C – collecting chamber; D – dialysis membrane; E – electrodes.

buffer, so that there is a continuous electrical path through the buffer from each electrode to the gel. This 'submarine' or 'submerged' gel system may not look as if it should work, as there is also a direct electrical path through the buffer from one electrode to the other, and molecules might be expected to diffuse from the gel into the overlying buffer. However, it is a system which does work very well, with the advantage that the gel is cooled by the buffer around it. Presumably sufficient current passes through the gel to move the DNA, and diffusion of DNA is so slow that little passes out of the gel during electrophoresis. A typical, home-made electrophoresis tank is shown in Figure 14.2. There are many variations on this theme; some designs leave out any raised support for the gel, which is placed on the bottom of a plastic box between two electrodes, and is then flooded with buffer.

Gel sizes depend on sample sizes and on the purpose of the electrophoresis. A general purpose gel could be 25 cm by 12 cm, resulting in a gel thickness of about 3 mm from 100 ml of gel solution. For rapid analysis of very small samples, it is common to use 'mini-gels', less than 10 cm long, which can be poured on small glass plates or on pieces of plastic or celluloid film cut to the size required.[4]

Sample Loading

With gel and buffer in place the samples are loaded using a microsyringe. For both vertical and horizontal gels the DNA samples must be dissolved in a solution containing sucrose, glycerol or Ficoll, to increase their density and so make them sink into the wells; usually a dye such as bromophenol blue is added to make it easier to follow the loading process and to act as a 'front' marker during electrophoresis. The amount of sample to be loaded will depend on the thickness of the gel, the complexity of the sample (i.e. how many different sizes of DNA fragments it contains), and the purpose of the gel (e.g. for molecular weight determinations, recovery of specific bands, autoradiography, or resolution of two similar fragments). It has been suggested that, for good resolution, up to 50 μg of a complex DNA mixture can be loaded for every cm^2 of gel the sample will run through.[5] However, it appears that loadings of more than 1.3 μg/cm^2 may cause distortion of bands and an increase in the mobilities of DNA molecules.[6] Clearly, in initial experiments it is advisable to compare the effects of different loadings of each sample, rather than relying on just one track. When the gel is being run for the determination of molecular weights from mobilities (see below), it is particularly important to avoid artificially high mobilities caused by overloading.

No 'stacking' gel is needed for the electrophoresis of DNA, as the mobilities of DNA molecules are much greater in the wells than in gel; consequently all the molecules 'pile-up' against the gel within a few minutes of turning on the current, forming a tight band at the start of the run.

Running the Gel

The voltage gradient used will depend on the sizes of molecules to be separated and on the agarose concentration. Large molecules diffuse very slowly in all except the most dilute gels, and so they are best separated using shallow voltage gradients for a long time; the low velocities of migration result in optimum resolution. However, small DNA fragments will diffuse more rapidly, causing broadening of bands during prolonged electrophoresis; so it is advisable to use much steeper voltage gradients for a short time to resolve them.

The concentration, and hence pore size, of the gel will obviously influence the time needed to resolve molecules of a certain size; and the efficiency of gel cooling will determine the maximum power which can be used. For the resolution of molecules between about 500 and 50,000 bp in a 0.8% agarose gel, a voltage gradient of about 1.5 V/cm, run overnight, is typical. When separating mixtures for the first time, or when using an unfamiliar buffer or gel system, it can be useful to examine the progress of the separation at intervals during the run. This can be done by including the dye ethidium bromide (see below) in the gel and running buffer, using ultraviolet light to show up the positions of the stained DNA (and RNA) bands in the gel at intervals during the run. This is made easier if the gel is poured on u.v.-transparent perspex instead of glass, so that the gel can remain on its base when being illuminated from below.

Detection of DNA

The reagent most widely used for revealing the positions of DNA bands on agarose (or polyacrylamide) gels is the fluorescent dye ethidium bromide. This intercalates into the DNA structure, and can be seen by its orange/red fluorescence when excited by u.v. light. The most efficient exciting wavelength is 254 nm, but this can seriously damage the DNA by nicking and dimerisation, and the dye is rapidly bleached. If DNA is to be recovered from the gel and restricted or cloned, photodamage can be a serious problem. The use of powerful u.v. of 300 nm provides little loss in sensitivity to ethidium bromide, yet greatly reduces the damage to DNA.[7] It should be remembered that u.v. light, especially at the intensities used for viewing gels, is harmful to unprotected eyes, so suitable goggles should always be worn.

Using 300 nm illumination from beneath the gel (transillumination) as little as 10 ng DNA can be seen in a band 1 cm wide. Photography of the gel, using a red filter to enhance contrast between the fluorescent bands and the background, may reveal bands which were too weak to be detected by eye, thus extending the sensitivity to nanogram levels of DNA. For wet processing, a film such as FP4 (Ilford) can be used. When prints are required immediately, Polaroid Type 57 or 667 film is suitable, and very high-quality negatives can be obtained using Types 55 or 665. It has been reported that a silver stain normally used for the ultrasensitive detection of proteins in polyacrylamide gels, will also reveal nanogram quantities of DNA and RNA in such gels, without the use of u.v. illumination.[8] This may prove to be a useful alternative to ethidium bromide when a transilluminator is not available, and when recovery of DNA from the gel is not necessary.

It is possible to quantify the amount of DNA in each band of a gel, by scanning densitometry of a photographic negative of the gel. However, corrections must be made for the sigmoid curve relating image density to the log of the incident light energy; but, with care, this method can be very accurate.[9]

If the DNA being separated is radioactive, the bands can be detected by autoradiography of the gel or of a filter after Southern blotting (Chapter 15).[10,11,12]

Measurement of Molecular Weights

The molecular weights of linear DNA molecules can be determined with considerable accuracy from their mobilities during gel electrophoresis. Possibly because of the analogy with SDS-PAGE of proteins, for which a plot of mobility vs log molecular weight produces a straight line (Chapter 2), it is common to do a similar plot for DNA gels. In fact, for double-stranded DNA in agarose gels, the relationship between mobility and log molecular weight is only approximately linear, and becomes markedly non-linear as the molecular weights or voltage gradient increase.[13] Duggleby *et al.*[14] have published a computer programme which fits a parabola to the results from a set of standards of known molecular weight, and which uses this parabola to determine the molecular weights of unknowns. This method is accurate to within 5%, but each parabola can only be applied over a narrow range of molecular weights. A more satisfactory approach has been suggested by Southern,[15,16] who showed that, if the length of a ds DNA molecule is L, and its mobility is m, then a plot of L vs $1/(m-m_o)$ is linear over a wide range of molecular weights, m_o being a correction factor, determined from standards. This linear relationship holds for ds DNA 0.2–6 kbp long in 2% agarose, 1–20 kbp in 1%, and 10–50 kbp in 0.5% agarose. By use of a computer and more size standards than the three used by Southern, Schaffer & Sederoff[17] have obtained accuracies better than 1% from L vs $1/(m-m_o)$ plots.

Recovery of DNA from Gels

For many applications it is sufficient to photograph the pattern of bands on a DNA gel, measure their mobilities, or carry out autoradiography, after which the gel is discarded. However, for such procedures as cloning or sequencing, the DNA of interest must be recovered from the gel. Normally a high yield of intact, undamaged molecules will be needed, and contaminants from the gel may have to be removed (agarose inhibits many enzymes used for restriction, ligation, etc.; non-crosslinked polyacrylamide seems to inhibit only reverse transcriptase but must be removed prior to electron microscopy of DNA).

Recovery methods have been reviewed by several authors.[18,19,20] The first step in each method is to locate the band of interest, usually using ethidium bromide, and then to separate the DNA from the gel. Owing to the slowness of diffusion of all except the smallest DNA molecules, the DNA must usually be helped out of the gel. This can be achieved by any of the following methods:

Electroelution.   A piece of gel containing the DNA band is cut out and transferred to dialysis tubing, which is filled with buffer and sealed. The tubing is then placed in an electroelution chamber, in which a voltage gradient is applied across it, causing the DNA to move out of the gel piece into the buffer within the dialysis tubing; the buffer can then be collected. Smith[21] gives details of a simple electroelution apparatus. DNA can be electroeluted *in situ* by placing a slip of DEAE-cellulose paper in a slit immediately to the anode side of the DNA band, and then continuing with the electrophoresis so that the DNA all moves to the paper strip, to which it will bind. Bound DNA is eluted from the paper using buffered NaCl, to give high recoveries of undamaged large DNA molecules;[22] hydroxyapatite can be used in a similar way to DEAE-cellulose paper.[23]

Gel Dissolution.   Agarose gels are held together by hydrogen-bonds, and so they will be dissolved by a chaotropic reagent such as KI. DNA can then be separated from agarose and KI using chromatography on hydroxyapatite,[24] or by isopycnic ultracentrifugation in a KI density gradient to remove agarose, followed by dialysis for elimination of KI.[25]

Using 6M $NaClO_4$ as chaotropic reagent, Chen and Thomas[26] were able to recover DNA in high yields by making use of the fact that it sticks to glass in such high concentrations of salt. The dissolved gel solution was passed through a glass-fibre filter, which bound the DNA, and the bound DNA was then washed free of agarose using more 6M $NaClO_4$. $NaClO_4$ was washed from the filter using ethanol, and the DNA was finally eluted from the glass-fibre by a dilute buffer. This method is rapid, does not damage DNA up to c. 10 kbp, and has the added advantage that ethidium bromide can be washed from DNA while it is attached to the filter.

Chaotropic reagents are unnecessary if low-melting-point agarose is used. Gel slices can be melted at 70°C, which is not hot enough to denature DNA; the DNA can then be complexed with quaternary ammonium cations, partitioned into 1-butanol, and finally partitioned back into aqueous solution as its sodium salt. The method works well for linear DNA up to about 25 kbp, but is unsuitable for supercoiled molecules larger than 16 kpb.[27]

Gel Compression.   Frozen gel slices can be crushed (a garlic press has been recommended) to extrude drops of buffer containing DNA,[28] especially after several cycles of freezing and thawing. Agarose contaminants can be removed chromatographically as described above. DNA recoveries vary from 50% to 90%.

Homogenisation and Diffusion.   To aid diffusion of large DNA molecules from the gel slice, it can be homogenised by being forced through a syringe needle prior to incubation in an eluting buffer. This method can work well for small DNA molecules, but not for large ones which either diffuse too slowly if homogenisation is too gentle, or are damaged by shearing during more violent homogenisation.[29]

Whichever approach is used to elute DNA from gels, it will usually be associated with ethidium bromide. This can be removed by repeated extraction with n-butanol or isoamyl alcohol, prior to further manipulation of the DNA.

5. Preparative Electrophoresis

The gel systems described so far are primarily analytical, although, as has been seen, microgram amounts of purified DNA can be recovered after excision of gel slices. Several preparative gel systems have been described, in which the separated DNA molecules are collected as they run off the anodic end of the gel during electrophoresis (Figure 14.3A). The designs differ in their method of collecting the eluted DNA and in their gel shape. The apparatus of Edgell and Polsky[30] is a logical extension of the familiar horizontal slab gel system. At the anode end of the gel, a dialysis membrane forms a partition between the anode chamber and a much smaller 'collecting' chamber (Figure 14.3B). Molecules running off the gel will be trapped in chamber C by the dialysis membrane, so will not be greatly diluted. Chamber C is regularly emptied and re-filled to recover the emerging DNA fractions. A gel 15 cm wide and 2 cm deep will handle up to 20 mg of restricted DNA, and by careful choice of sampling interval the restriction fragments of an EcoRI digest of λDNA can be resolved. Southern[31] has designed an annular gel, in which sample is loaded at the perimeter and electrophoresed to the centre, thus helping to concentrate the eluted fractions. Other preparative gels are designed for continuous elution of the molecules running off the gel. The success of these gels depends on the minimisation of volume in the 'collecting chamber', and an elution buffer flow rate which is fast enough to resolve different bands, but not so fast that each band is greatly diluted. One such system is produced by Bethesda Research Laboratories, Inc., and uses sintered glass to produce a low volume 'collecting chamber' in contact with the anode end of a rod gel. Up to 0.5 mg DNA can be run on a 1 cm diameter gel, and resolution is comparable to that of the Edgell and Polsky[32] system.

6. Alternatives to Agarose Gel Electrophoresis

Polyacrylamide Gel Electrophoresis

In spite of the versatility of agarose gel electrophoresis, there are times when alternative separation methods are preferable. For the separation of molecules shorter than about 1000 bp, polyacrylamide gel electrophoresis is particularly useful. Maniatis *et al.*[33] used 12% gels to resolve fragments between 20 and 100 bp, and 3.5% gels for the range 80 to 1000 bp. Under these conditions there was a linear relationship between mobility and log molecular weight. Single-stranded DNA of the same size range can be successfully fractionated using denaturing gels to prevent the formation of duplexes, and this technique has been reviewed by Maniatis and Efstratiadis.[34] Short DNA molecules remain denatured in the presence of 7M urea, but 98% formamide is needed to keep molecules between 100–1000 nt in their single-stranded form. A method using glyoxal as denaturing reagent is described in Chapter 6.

Sometimes the fragments of DNA produced using a restriction endonuclease have such a wide range of sizes that they cannot all be resolved on a single conventional gel. In such cases it may be worth using a polyacrylamide gradient gel for electrophoresis.[35] A typical gradient gel ranges from 3.5% acrylamide at the top to 7.5% at its base, the concentration changing smoothly along the gel. In such a gel the DNA molecules experience increasing resistance to their movement as they move downwards into more concentrated gel, and this results in a continual sharpening of each band of DNA, as the trailing molecules tend to have higher mobilities than the leading ones. It is hardly surprising that the relationship between molecular weight and mobility is rather complex in these gels: a plot of log molecular weight against log mobility produces an approximation to a sigmoid curve. Clearly this is not a good system for accurate molecular weight measurements, but it is of value in resolving a large number of DNA molecules of widely varying sizes on one gel.

DNA can be recovered from agarose by the use of chaotropic reagents to dissolve the gel; however, polyacrylamide gels are covalently cross-linked and therefore resist solubilisation. Hansen[36] has found a way round this limitation by using bis-acrylylcystamine (BAC) in place of bis-acrylamide when preparing polyacrylamide gels; in this case the gel is held together by disulphide cross-linkages, which can be broken using mercaptoethanol. A 5% BAC-acrylamide gel was found to give excellent resolution with DNA up to 6 kbp, and after electrophoresis it was easily solubilised using mercaptoethanol. DNA can be recovered from the gel solution by chromatography on DEAE-cellulose, after which it is suitable for further manipulation.

Column Chromatography

Pearson *et al.*[37] described the use of a reversed-phase chromatography (RPC) column, RPC-5, for separation of tRNAs. This type of column is now being used successfully for the preparative separation of DNA restriction fragments,[38] and appears to be a very powerful method for DNA analysis. For more details, the reader is referred to Chapter 1.

References

1. E.M. Southern, 'Measurement of DNA length by gel electrophoresis', *Anal. Biochem.*, *100* (1979), 319–23.
2. A.N. Best, D.P. Allison and G.D. Novelli, 'Purification of supercoiled DNA of plasmid ColE1 by RPC-5 chromatography', *Anal. Biochem.*, *114* (1981), 235–43.
3. P.H. Johnson, M.J. Miller and L.I. Grossman, 'Electrophoresis of DNA in agarose gels. II. Effects of loading mass and electroendosmosis on electrophoretic mobilities', *Anal. Biochem.*, *102* (1980), 159–62.
4. J.J. Kopchick, B.R. Cullen and D.W. Stacey, 'Rapid analysis of small nucleic acid samples by gel electrophoresis', *Anal. Biochem.*, *115* (1981), 419–23.
5. E.M. Southern, 'Gel electrophoresis of restriction fragments', *Methods in Enzymology*, *68* (1979), 152–76.
6. Johnson, Miller and Grossman, 'Electrophoresis of DNA in agarose gels'.
7. C.F. Brunk and L. Simpson, 'Comparison of various ultraviolet sources for fluorescent detection of ethidium bromide-DNA complexes in polyacrylamide gels', *Anal. Biochem.*, *82* (1977), 455–62.
8. L.L. Somerville and K. Wang, 'The ultrasensitive silver "protein" stain also detects nanograms of nucleic acids', *Biochem. Biophys. Res. Comm.*, *102* (1981), 53–8.
9. A. Prunell, 'A photographic method to quantitate DNA in gel electrophoresis', *Methods in Enzymology*, *65* (1980), 353–8.
10. R.A. Laskey and A.D. Mills, 'Quantitative film detection of ^{3}H and ^{14}C in polyacrylamide gels by fluorography', *Eur. J. Biochem.*, *56* (1975), 335–41.
11. R.A. Laskey and A.D. Mills, 'Enhanced autoradiographic detection of ^{32}P and ^{125}I using intensifying screens and hypersensitized film', *FEBS Letts.*, *82* (1977), 314–16.
12. R.A. Laskey, 'The use of intensifying screens or organic scintillators for visualizing radioactive molecules resolved by gel electrophoresis', *Methods in Enzymology*, *65* (1980), 363–71.
13. Southern, 'Measurement of DNA length'.
14. R.G. Duggleby, H. Kinns and J.I. Rood, 'A computer program for determining the size of DNA restriction fragments', *Anal. Biochem.*, *110* (1981), 49–55.
15. Southern, 'Measurement of DNA length'.
16. Southern 'Gel electrophoresis'.
17. H.E. Schaffer and R.R. Sederoff, 'Improved estimation of DNA fragment lengths from agarose gels', *Anal. Biochem.*, *115* (1981), 113–22.
18. R.C.-A. Yang, J. Lis and R. Wu, 'Elution of DNA from agarose gels after electrophoresis', *Methods in Enzymology*, *68* (1979), 176–82.
19. H.O. Smith, 'Recovery of DNA from gels', *Methods in Enzymology*, *65* (1980), 371–80.
20. C.W. Chen and C.A.Jr. Thomas, 'Recovery of DNA segments from agarose gels', *Anal. Biochem.*, *101* (1980), 339–41.
21. Smith, 'Recovery of DNA from gels'.
22. G. Dretzen, M. Bellard, P. Sassone-Corsi and P. Chambon, 'A reliable method for the recovery of DNA fragments from agarose and acrylamide gels', *Anal. Biochem.*, *112* (1981), 295–8.
23. Southern, 'Gel electrophoresis'.
24. F.C. Wheeler, R.A. Fishel and R.C. Warner, 'Agarose gel electrophoresis of circular DNA of replicative form of bacteriophage G4', *Anal. Biochem.*, *78* (1977), 260–75.
25. Smith, 'Recovery of DNA from gels'.
26. Chen and Thomas, 'Recovery of DNA segments'.

27. J. Langridge, P. Langridge and P.L. Bergquist, 'Extraction of nucleic acids from agarose gels', *Anal. Biochem., 103* (1980), 264–271.
28. Wheeler, Fishel and Warner, 'Agarose gel electrophoresis'.
29. Ibid.
30. M.H. Edgell and F.I. Polsky, 'Use of preparative gel electrophoresis for DNA fragment isolation', *Methods in Enzymology, 65* (1980), 319–27.
31. E.M. Southern, 'A preparative gel electrophoresis apparatus for large scale separations', *Anal. Biochem., 100* (1979), 304–18.
32. Edgell and Polsky, 'Use of preparative gel electrophoresis'.
33. T. Maniatis, A. Jeffrey and J.H. van de Sande, 'Chain length determination of small double- and single-stranded DNA molecules by polyacrylamide gel electrophoresis', *Biochemistry, 14* (1975), 3787–94.
34. T. Maniatis and A. Efstratiadis, 'Fractionation of low molecular weight DNA or RNA in polyacrylamide gels containing 98% formamide or 7M urea', *Methods in Enzymology, 65* (1980), 299–305.
35. P.G.N. Jeppesen, 'Separation and isolation of DNA fragments using linear polyacrylamide gradient gel electrophoresis', *Methods in Enzymology, 65* (1980), 305–19.
36. J.N. Hansen, 'Use of solubilizable acrylamide disulfide gels for isolation of DNA fragments suitable for sequence analysis', *Anal. Biochem., 116* (1981), 146–151.
37. R.L. Pearson, J.F. Weiss and A.D. Kelmers, 'Improved separation of transfer RNAs on polychlorotrifluoroethylene-supported reversed-phase chromatography columns', *Biochim. Biophys. Acta, 228* (1971), 770–4.
38. R.D. Wells, S.C. Hardies, G.T. Horn, B. Klein, J.E. Larson, S.K. Neuendorf, N. Panayotatos, R.K. Patient and E. Selsing, 'RPC-5 column chromatography for the isolation of DNA fragments', *Methods in Enzymology, 65* (1980), 327–47.

15 DETECTION OF SPECIFIC DNA SEQUENCES – THE SOUTHERN BLOT

Christopher G.P. Mathew

CONTENTS

1. Introduction

In a previous chapter (Chapter 14) the characterisation of DNA by restriction analysis on agarose gels has been described. In such cases the DNA sample would normally be about one microgram of a single type of DNA sequence. However, if you wished to analyse the structure of a specific gene from a complex organism without prior purification of the gene, you would have to detect picogram amounts of a single type of DNA sequence among about a million other types. The method used would therefore have to be very sensitive and very specific. This exacting task can be accomplished by means of a technique devised by Edwin Southern of Edinburgh University.[1] The technique has come to be known as the Southern transfer or Southern blot, and has had an enormous impact on molecular biology and genetics.

The initial step in the procedure is to digest the DNA with a restriction enzyme, and to separate the resulting fragments by electrophoresis on an agarose gel. The DNA is then denatured and transferred to a nitrocellulose filter, which is incubated with a radioactive RNA or DNA 'probe' for a particular gene. The probe will then seek out complementary DNA sequences on the filter, and associate with them by hydrogen bonding between the bases on the respective strands. After washing off unbound probe, an x-ray film is placed in contact with the filter and exposed. The position of DNA fragments on the filter that contain the gene of interest will then be detected as dark bands on the developed film.

2. Experimental Outline

A diagram illustrating all the steps in the procedure is given in Figure 15.1.

Restriction and Electrophoresis of DNA

Restriction endonuclease digestion of the DNA is normally carried out at 37°C in the buffer specified for that enzyme by the manufacturer. One unit of enzyme activity is usually defined as the amount required to digest completely one microgram of λ DNA in one hour at 37°C, although in practice, a complex, high molecular weight DNA is digested with an excess of enzyme for several hours. Bovine serum albumin can be added to the buffer to enhance enzyme stability. The completeness of digestion of the

Figure 15.1: Diagrammatic Representation of the Steps involved in the Detection of a Hypothetical β-globin gene

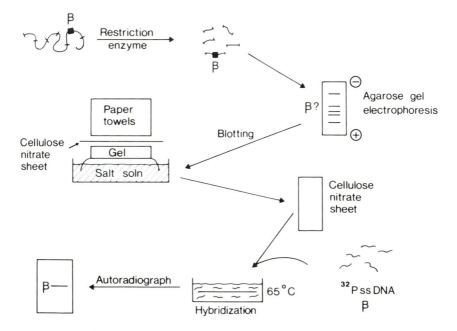

Source: P.F.R. Little, 'Antenatal diagnosis of haemoglobinopathies' in R. Williamson (ed.), *Genetic Engineering* (Academic Press, 1981), vol. 1, 61–102. (With modifications)

DNA should be ascertained by adding a simple DNA (e.g. λ DNA) to an aliquot of the digest, and running this on a checking gel before electrophoresis of the entire sample.

Details of DNA analysis by gel electrophoresis have been discussed in Chapter 14. Agarose gels are normally used for Southern blots, but if small DNA fragments are to be resolved (< 250 bp), composite gels of agarose and polyacrylamide may be used.[2] Molecular weight markers and hybridisation markers (20–100 pg of unlabelled probe) should be electrophoresed with the samples. Gels are run overnight at low voltage for optimum resolution, then stained with ethidium bromide and photographed (see Figure 15.2A).

Transfer of DNA to Nitrocellulose Filters

The object of this procedure is to transfer all the DNA from the agarose gel onto a nitrocellulose filter so that the original pattern of fragments is preserved.[1] Since double-stranded DNA is not retained by nitrocellulose, the DNA is first denatured by soaking the gel in sodium hydroxide. A sheet

Figure 15.2: Detection of Specific Sequences by Blot Hybridisation

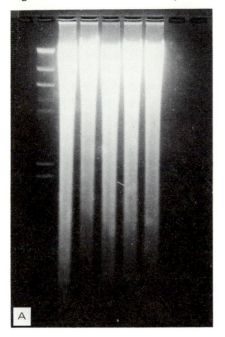
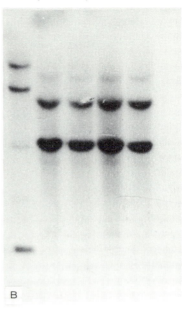

Note A: Human DNA digested with restriction enzyme and electrophoresed on a 0.8% agarose gel. The first lane contains a λ DNA molecular weight marker. Note the continuous smear of human DNA fragments in the other lanes.

Note B: Autoradiograph of samples from Figure 15.2A after blotting and hybridisation with a β-globin probe. Only DNA fragments which contain β-globin sequences are detected.

of nitrocellulose paper is placed over the gel, and solvent blotted through it by stacking absorbent paper on top (see Figure 15.1). The DNA is carried out of the gel by the flow of solvent and trapped in the nitrocellulose paper. After baking the DNA onto the filter (two hours at 80°C), the filter may be stored for several months at 4°C.

Several factors affect the efficiency of the transfer procedure:

(i) Transfer of large DNA fragments. DNA fragments larger than 10 Kb in size are transferred very slowly, and should be broken down in the gel prior to transfer. This can be done either by irradiating the gel with short-wave ultraviolet rays before denaturation,[4] or by partial depurination with dilute acid followed by strand cleavage with alkali.[5] The effectiveness of the irradiation method has been disputed.[6]

(ii) Choice of nitrocellulose paper. Some brands of nitrocellulose seem to retain DNA more efficiently than others. We have found that the sensitivity of detection of specific sequences is several-fold higher if

Schleicher and Schuell rather than Millipore nitrocellulose paper is used (see also Faub *et al.*[7]). Different pore sizes of these filters are available – the standard size is 0.45 μm, but the 0.2 μm size is apparently far superior for binding small DNA fragments.[8] For transfer of DNA fragments less than about 250 bp in length, alternative transfer media should be used (see Section 4).

(iii) Time of transfer. This will depend on the thickness of the gel, the percentage agarose in the gel, and the sizes of the fragments to be transferred. An overnight transfer is usually sufficient, especially if the DNA has been fragmented in the gel as described above. Recently, an electrophoretic method of transfer has been developed,[9] which reduces transfer time to as little as one hour. This would be the method of choice if the DNA was to be transferred from a polyacrylamide gel, since blotting from polyacrylamide is very slow, and can only be effected if a cleavable crosslinking agent is used for gel formation.[10]

Hybridisation of Probe to Filter

Before the filter can be incubated with a radioactive probe it is washed several times in a solution of high molecular weight polymers devised by Denhardt.[11] This prevents further binding of single-stranded DNA to the filter, but does not affect specific annealing of complementary sequences.

The filters are then usually sealed in a polythene bag with the solution containing the probe, and incubated for 1–48 hours. Unless heterologous DNA is added to the hybridisation solution, a small but significant amount of labelled DNA may stick to all the DNA on the filter, and produce a high background on the autoradiograph. Conditions for optimal signal/noise ratios have been described.[12] The addition of dextran sulphate to the hybridisation solution has been reported to increase the rate of hybridisation substantially.[13]

Many methods are available for radiolabelling the sequence probe. ^{3}H or ^{32}P-cDNA, nick translated DNA, RNA terminally labelled with ^{32}P or ^{125}I-labelled RNA have all been used. The nick translation procedure is described in Chapter 8.

After incubation with the probe, the filters are washed in solutions of low salt concentration in order to remove unbound probe. The lower the salt concentration of the wash, the lower the background of non-specifically bound DNA is likely to be. However, low salt concentration washes should only be used when the sequences of the probe and the filter-bound DNA are very closely matched (see also Section 3).

Autoradiography

Once the nitrocellulose filter has been washed and dried it is put into a light-tight cassette, an x-ray film is placed over it, followed by an intensifying screen, and the loaded cassette kept at –70°C. The time of exposure required to detect the specific sequences will obviously depend on the amount of specific sequence DNA on the filter and the specific activity of the probe. In order to detect a unique sequence (one copy per haploid set of chromosomes), an exposure of several days may be necessary. The x-ray film is then developed and fixed (see Figure 15.2B).

The sensitivity of the autoradiography is dependent on several elements:

(i) The x-ray film should be hypersensitised by pre-exposure to an instantaneous flash of light from an electronic photographic flash unit. This is also important if the relative intensities of the bands are to be quantified.[14]

(ii) The sensitivity of the various commercially available x-ray films varies quite widely. Kodak X – Omat R and Fuji RX are suitable.[15]

(iii) The relative merits of alternative types of intensifying screens have been investigated.[16] Du pont Cronex lightning plus and Fuji Mach 2 calcium tungstate screens were found to be the most sensitive.

(iv) The film should be exposed at –70°C for optimal sensitivity and for accurate quantitative work. However, pre-exposure and low temperature exposure will only increase sensitivity if an intensifying screen is used.

(v) If the probe used in the hybridisation is labelled with a low energy β-emitter such as tritium, the filter must be impregnated with an organic scintillator such as PPO (2,5-diphenyloxazole) to allow detection of the signal. A spray-on scintillator is commercially available (Enhance, New England Nuclear).

The relative intensity of the bands on the autoradiograph, and hence the relative amounts of the sequences they represent, may be assessed quantitatively by scanning the developed film. However, inherent technical problems such as incomplete transfer of large fragments and low efficiency of hybridisation of small fragments must be taken into account.[17]

Practicability

The Southern transfer is a rather lengthy experimental procedure, which will take about seven to ten days from restriction of the DNA to development of the autoradiograph. However, many samples can be analysed simultaneously, and much of the time is 'passive' – e.g. leaving filters to incubate. No highly specialised equipment is necessary other than that required for electrophoresis (Chapter 14). The costly aspects of the procedure are the purchase of restriction enzymes, and of ^{32}P-containing nucleotides for radioactive labelling of the probe.

3. Theoretical Aspects

Although the general outline of a Southern transfer experiment will not vary from that described in Section 2, some important experimental details may be altered to suit the needs of the investigator. It is as well to consider some theoretical aspects of the procedure, so that the basis of these alterations can be understood.

The Southern transfer technique depends upon the fact that single-stranded DNA is absorbed by nitrocellulose filters. The nature of this binding is poorly understood, but it is known to be weakened by a rise in temperature, and strengthened by a rise in salt concentration.[18] The transfer itself is carried out at a very high salt concentration, and a relatively high concentration should be maintained throughout the subsequent hybridisation and washing steps if binding is to be encouraged. The length of DNA strand is also important for binding, since small DNA fragments (<250 bp) are not bound efficiently.[19]

The rate and precision of probe hybridisation to complementary sequences on the filter depend on several factors. These include:

(i) Probe concentration. Hybridisation will clearly be more rapid at high probe concentrations, since the possibility of two complementary strands encountering each other will be higher.

(ii) Sequence complexity of the probe. If the probe is a simple sequence repeated many times in the DNA of the genome, the effective concentration of complementary sequences will be high and hence also the rate of hybridisation. Unique (single copy) sequences will take a lot longer to hybridise to detectable levels.

(iii) Temperature. This provides a means of controlling the precision or stringency of hybridisation. The higher the temperature, the more precisely matched two sequences must be to anneal with each other. At lower temperatures, duplexes can be formed with more mismatching. Efficiency of reassociation of unique sequences is maximal at 20°C below the melting temperature, i.e. at about 65°C. Hybridisation and washing of filters is normally done at this temperature.

(iv) Solvent. The melting temperature of DNA can be reduced by organic solvents which assist denaturation. In practice this means that the inclusion of solvents such as formamide in the hybridisation solution enables one to perform stringent hybridisations at a temperature of about 40°C.

(v) Salt concentration. A high ionic strength will promote hybridisation since the mutual repulsion of two highly charged DNA molecules can be masked by the presence of salt. This effect is also of great importance when considering the ionic strength of the solution with which unbound probe is to be washed from the filter after hybridisation has been completed. If a low ionic strength wash is used, the probe will only remain bound to sequences of very high complementarity. In an experiment where, for example, a chicken-derived probe is being used to detect a human gene, a series of different ionic strength washes should be tried in order to establish the optimal salt concentration.

In the final step of the procedure, i.e. detection of bound radioactive probe by autoradiography, the main issue is how to improve detection efficiency. If the probe has been labelled with ^{32}P-containing nucleotides, then the high-energy β particles will expose the x-ray film directly. However, the sensitivity of detection of these particles can be greatly enhanced by solid-state scintillation.[20] This means placing a high-density fluorescent 'intensifying screen' on top of the x-ray film so that the excess energy of the β particles emitted from the filter will be captured by the screen and returned to the film in the form of visible light. For maximum efficiency of detection and for quantification of the image, the initial log phase of image formation in the film must be by-passed. This is achieved by pre-exposing the film to an instantaneous flash of light.[21] If the probe has been labelled with a weak β-emitter such as tritium, the nitrocellulose filter must be impregnated with an organic scintillator. This will convert the emitted energy to visible light, and thus increase the penetration range of the β-particles.

4. Alternative Transfer Media

The use of nitrocellulose filters for detection of specific sequences has two serious limitations; neither RNA nor small DNA fragments (< 250 bp) bind efficiently to nitrocellulose. These difficulties have been overcome by linking RNA or DNA covalently to chemically-activated paper. Aminobenzyloxymethyl (ABM) paper is activated to the diazo form (DBM) immediately before use. The diazonium group then reacts with single-stranded nucleic acids to form covalent azo derivatives (Figure 15.3). Nitrobenzyloxymethyl (NBM) or aminobenzyloxymethyl (ABM) paper can be purchased, and activated to the diazo form (DBM) immediately before use.

Figure 15.3: Covalent Linkage of Nucleic acid to DBM Paper

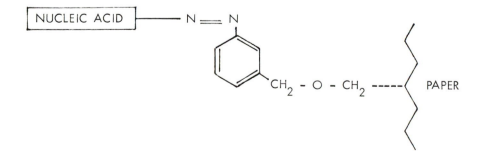

DNA fragments as small as 30 nucleotides long have been transferred and detected using DBM paper.[22] Another considerable advantage is that the covalent linkage allows hybridised probe to be 'scrubbed off', and the transfer re-hybridised to a different probe. Aminophenylthioether (APT) derivatised paper has also been used for this purpose.[23,24]

Transfer of RNA to DBM paper has been dubbed 'Northern transfer' to distinguish the procedure from transfer of DNA to nitrocellulose.

5. Applications

A detailed list of the many uses to which the Southern transfer has been put is beyond the scope of this chapter. The following discussion should give some idea of the range of applications of this powerful technique.

Analysis of Gene Structure and Expression

In bacteria, genes consist of a contiguous stretch of DNA, and the linear sequence of nucleotides corresponds directly to a linear sequence of amino acids in the protein. The discovery that this principle of colinearity did not exist in the genomes of higher eukaryotes was an event of profound significance in molecular biology, and was made by Southern blot analysis of genomic DNA. Chambon and his colleagues, for example, used complementary DNA (cDNA) made from ovalbumin mRNA to probe ovalbumin gene structure in the chicken.[25] Although ovalbumin cDNA was not cut by the restriction enzymes *Eco* Rl or *Hind* 111, restriction of chicken genomic DNA with one or other of these enzymes, followed by blotting and hybridisation with an ovalbumin cDNA probe, showed several bands on the autoradiograph. Thus the ovalbumin gene had to be interrupted by other DNA sequences not represented in the ovalbumin mRNA. This was soon shown to be a general phenomenon in higher eukaryotic genomes.

The technique has also proved useful in analysing the role of DNA methylation in gene expression. A proportion of cytosine residues in the DNA of higher eukaryotes is methylated, and the extent of methylation seems to be inversely related to transcriptional activity (reviewed by Ehrlich and Wang). A gene will therefore be less methylated in a tissue in which it is expressed than in a tissue in which it is inactive. The level of gene methylation can be investigated by Southern blot analysis using the restriction enzymes *Mspl* and *Hpa* ll. Both of these enzymes cut at the sequence CCGG, but *Hpa* ll will not cut if the internal cytosine of this sequence is methylated i.e. Cm^5CGG, whereas *Msp* l will. Thus a comparison of the *Hpa* ll and *Msp* l restriction patterns of a gene will reflect the extent to which it is methylated.

Studies of gene activation have also benefitted from this technique. It is known that genes which are being transcribed in a particular tissue are more rapidly degraded by DNAse l than inactive genes. The Southern blot is the basis of a very sensitive assay for this preferential degradation.[27] Nuclei are digested with DNAse 1, and the DNA extracted and cut with a restriction enzyme. The DNA is then electrophoresed, blotted, and probed for a particular gene. The disappearance of specific restriction fragments containing the gene as a result of their preliminary degradation by DNAse 1 will then reflect the sensitivity of the gene to this enzyme. This assay can be used to compare the activity of a particular gene at different stages of development or in the different tissues of an organism.

Molecular Genetics

The Southern blot was originally developed to study the gene coding for ribosomal RNA,[28] but it was soon applied to the analysis of malfunctioning genes in various inherited diseases. In particular, disorders of globin gene expression such as thalassaemia and sickle cell anaemia have been examined in great detail, since sequence probes for these genes are available. Restriction mapping of the β-globin gene in patients with β-thalassaemia (decreased β-globin chain production) has revealed the gene to be almost always structurally intact, whereas the α-globin genes are usually deleted in α-thalassaemia.[29] It has even been possible to detect the point mutation in the 1600 nucleotide β-globin gene which gives rise to sickle cell anaemia.[30] In this condition a glutamic acid residue is replaced by a valine as a result of the nucleotide sequence mutation $GAG \rightarrow GTG$. It so happens that the restriction enzyme Dde 1 cuts at the sequence CTGAG, so that this cutting site is destroyed in the sickle β-globin gene. Blot hybridisation analysis therefore detects a 376 base pair band instead of two bands of 201 and 175 base pairs. Furthermore, since fetal DNA can be obtained from amniotic fluid at about the 16th week of pregnancy, blot hybridisation can be used for antenatal diagnosis of this condition.

If the mutation which causes an inherited disorder cannot be detected directly with one of the known restriction enzymes, it can often be detected indirectly by means of restriction enzyme polymorphisms. In this approach an unimportant alteration in DNA sequence near the gene in question, which creates or abolishes a restriction enzyme cutting site, can be used to trace the affected gene in a particular family or population. This was first demonstrated by Kan and Dozy[31] who used a polymorphism for the restriction enzyme *Hpa* 1 for the antenatal diagnosis of sickle cell anaemia. They found that the β-globin genes of most of their black patients with sickle cell disease were located on a 13 Kb *Hpa* 1 fragment, whereas the normal β-globin genes were usually on a 7.6 Kb fragment. Such polymorphisms should eventually permit diagnosis of any inherited disorder caused by a defect in a single gene.[32]

The existence of restriction fragment polymorphisms has also been exploited in the field of population genetics and evolution. For example, variations in the *Hpa* 1 cleavage pattern of human mitochondrial DNA of various population groups have been used to support the theory that Asia is genetically central to the migrations that gave rise to the human ethnic groups.[33]

Further Reading

Note: details of current methodology will be found in references 2, 4, 5, 12 and 14.
1. P. Chambon, 'Split genes', *Sci. Amer.*, *244* (1981), 48–59
2. B.D. Hall, L. Haar and K. Klepp, 'Development of the nitrocellulose filter technique for RNA-DNA hybridisation', *Trends in Biochemical Sciences*, 5 (1980), 254–6
3. P.F.R. Little, 'Ante-natel diagnosis of haemoglobinopathies' in R. Williamson (ed.), *Genetic Engineering*, (Academic Press, 1981) vol. 1, 61–102.

References

1. E.M. Southern, 'Detection of Specific Sequences among DNA fragments separated by gel electrophoresis', *J. Mol. Biol.*, *98* (1975), 503–17.
2. J.C. Alwine, D.J. Kemp, B.A. Parker, J. Reiser, J. Renart, G.R. Stark and G.M. Wahl, 'Detection of specific RNAs or specific fragments of DNA by fractionation in gels and transfer to diazobenzyloxymethyl paper' in R. Wu, (ed.), *Methods in Enzymology*, *68* (1979), 220–42.
3. Southern, 'Detection of specific sequences among DNA fragments'.
4. E.M. Southern, 'Gel electrophoresis of restriction fragments' in R. Wu, (ed.), *Methods in Enzymology*, *68* (1979), 152–76.
5. G.M. Wahl, M. Stern and G.R. Stark, 'Efficient transfer of large DNA fragments from agarose gels to DBM paper and rapid hybridisation using dextran sulphate', *Proc. Natl. Aca. Sci. USA*, *76* (1979), 3683–7.
6. M. Bittner, P. Kupferer and C.F. Morris, 'Electrophoretic transfer of proteins and nucleic acids from slab gels to diazobenzyloxymethyl cellulose or nitrocellulose sheets', *Anal. Biochem.*, *102* (1980), 459–71.
7. O. Faub, S. Bratosin, M. Horowitz and Y. Alom, 'The initiation of transcription of SV40 DNA at late time after infection', *Virology*, *92* (1979), 310–23.
8. Ibid.
9. Bittner, Kupferer and Morris, 'Electrophoretic transfer'.
10. Alwine, Kemp, Parker, Reiser, Renart, Stark and Wahl, 'Detection of specific RNAs'.
11. D.T. Denhardt, 'A membrane-filter technique for the detection of complementary DNA', *Biochem. Biophys. Res. Comm.*, *23* (1966), 641–6.
12. M. Barinaga, R. Franco, J. Meinkoth, E. Ong and G.M. Wahl, (1981) 'Methods for transfer of DNA, RNA and protein to nitrocellulose paper and diazotized paper solid supports'. Schleicher and Schuell technical publication No. 352–4.
13. Wahl, Stern and Stark, 'Efficient transfer of large DNA fragments'.
14. R.A. Laskey, 'The use of intensifying screens or organic scintillators for visualizing radioactive molecules resolved by gel electrophoresis' in L. Grossman and K. Moldave (eds.), *Methods in Enzymology*, *65* (1980), part 1, 363–71.
15. R.A. Laskey and A.D. Mills, 'Enhanced autoradiographic detection of ^{32}P and ^{125}I using intensifying screens and hypersensitized film', *FEBS Letts*, *82* (1977), 314–8.
16. Ibid.
17. Southern, 'Gel electrophoresis'.
18. D. Gillespie, 'The formation and detection of DNA-RNA hybrids' in L. Grossman and K. Moldave (eds.), *Methods in Enzymology*, *12B* (1968), 641–68.
19. Barinaga, Franco, Meinkoth, Ong and Wahl, 'Methods for transfer'.
20. Laskey, 'The use of intensifying screens'.
21. Ibid.
22. Alwine, Kemp, Parker, Reiser, Renart, Stark and Wahl, 'Detection of specific RNAs'.
23. Ibid.
24. Barinaga, Franco, Meinkoth, Ong and Wahl, 'Methods for transfer'.

25. R. Breathnach, J.L. Mandel, P. Chambon, 'Ovalbumin gene is split in chicken DNA', *Nature*, *270* (1977), 314–19.
26. M. Ehrlich and R. Wang, '5–Methylcytosine in eukaryotic DNA', *Science*, *212* (1981), 1350–7.
27. J. Stalder, M. Groudine, J.B. Dodgson, J.D. Engel and H. Weintraub, 'Hb switching in chicken', *Cell*, *19* (1980), 973–80.
28. Southern, 'Detection of specific sequences among DNA fragments'.
29. T. Maniatis, E. Fritsch, J. Lauer and R. Lawn, 'The molecular genetics of human haemoglobins', *Ann. Rev. Genet.*, *14* (1980), 145–78.
30. R.T. Geever, L.B. Wilson, F.S. Nallaseth, P.F. Milner, M. Bittner and J.T. Wilson, 'Direct identification of sickle cell anaemia by blot hybridisation', *Proc. Natl. Acad. Sci. USA*, *78* (1981), 5081–5.
31. Y.W. Kan and A.M. Dozy, 'Ante-natal diagnosis of sickle cell anaemia by DNA analysis of amniotic fluid cells', *The Lancet*, *ii* (1978), 910–2.
32. D. Botstein, R.L. White, M. Skolnick and R. Davis, 'Construction of a genetic linkage map in man using restriction fragment length polymorphisms', *Amer. J. Hum. Genet.*, *32* (1980), 314–31.
33. M. Denaro, H. Blanc, M. Johnson, K. Chen, E. Wilmsen, L. Cavalli-Sforza and D. Wallace, 'Ethnic variation in Hpa 1 endonuclease cleavage patterns of human mitochondrial DNA', *Proc. Natl. Acad. Sci. USA.*, *78* (1981) 5768–2.

16 THE DETERMINATION OF DNA SEQUENCES

Wim Gaastra and Bauke Oudega

CONTENTS

1. Introduction

Unlike the determination of the amino acid sequence of proteins and peptides, which is based on the sequential degradation of these structures and the subsequent identification of the cleaved-off amino acid residue, DNA sequence analysis is based on high-resolution electrophoresis on denaturing polyacrylamide gels of oligonucleotides with one common end, and varying in length by a single nucleotide at the other end. Although there are several rapid methods available today for DNA sequence analysis, they all rely on high-resolution gel electrophoresis. The main difference between the methods currently used lies in the way in which the set of oligonucleotides to be separated are produced. There are two approaches to obtaining such a set of oligonucleotides. Sanger and his collaborates have developed various enzymatic methods for the determination of the sequence of nucleotides in DNA.[1,2] In their procedure a DNA strand to be sequenced is used as a template for *E. coli* DNA polymerase I, and a short complementary fragment is used as a primer. The primer is annealed to the template and then extended enzymatically for an average of 15 to 300 or more nucleotides in the presence of radioactive labelled deoxyribonucleoside triphosphates.

If this primed synthesis is performed under conditions where the synthesis is ended at a specific nucleotide, a set of oligonucleotides can be obtained, all ending with the same nucleotide. The four sets of oligonucleotides obtained in this way, one for each nucleotide present in DNA, are then run in parallel on the same gel and after autoradiography the sequence can be read (see Figure 16.3).

In the Maxam and Gilbert method for sequencing DNA, the four sets of oligonucleotides are obtained by treating a ^{32}P-endlabelled DNA fragment under four different conditions with a reagent that cleaves next to a particular nucleotide,[3] after which gel electrophoresis and autoradiography are performed in the same way as above (see Figure 16.5).

2. Sequencing DNA by Primed Synthesis

The 'Plus and Minus Method'

In the 'minus' method, DNA polymerase I is first used for the primed synthesis of a radioactive complementary copy of a single-stranded

Figure 16.1 The Principle of the Plus and Minus Methods for Determination of the Nucleotide Sequence in DNA

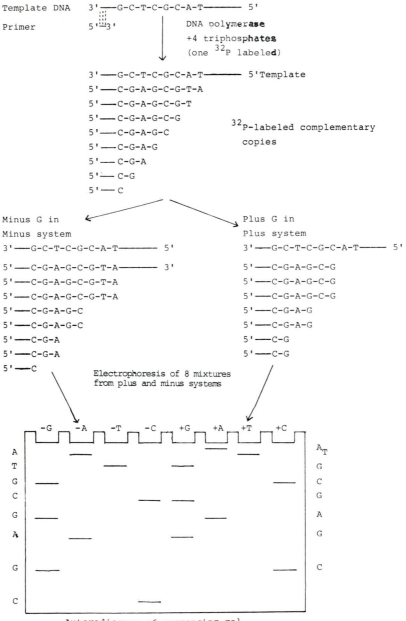

```
Template DNA    3'——G-C-T-C-G-C-A-T———— 5'
                      ┊┊┊
Primer          5'┈┈3'              DNA polymerase
                                    +4 triphosphates
                                    (one ³²P labeled)

                3'——G-C-T-C-G-C-A-T———— 5'Template
                5'——C-G-A-G-C-G-T-A
                5'——C-G-A-G-C-G-T
                5'——C-G-A-G-C-G                ³²P-labeled complementary
                5'——C-G-A-G-C                     copies
                5'——C-G-A-G
                5'——C-G-A
                5'——C-G
                5'——C
```

Minus G in   Plus G in
Minus system   Plus system
```
3'——G-C-T-C-G-C-A-T———— 5'              3'——G-C-T-C-G-C-A-T———— 5'

5'——C-G-A-G-C-G-T-A———— 3'              5'——C-G-A-G-C-G
5'——C-G-A-G-C-G-T-A                     5'——C-G-A-G-C-G
5'——C-G-A-G-C-G-T-A                     5'——C-G-A-G-C-G
5'——C-G-A-G-C                           5'——C-G-A-G
5'——C-G-A-G-C                           5'——C-G-A-G
5'——C-G-A                               5'——C-G
5'——C-G-A                               5'——C-G
5'——C
```

Electrophoresis of 8 mixtures
from plus and minus systems

Autoradiogram of sequencing gel

Note: The sequence is read from the bottom to the top of the gel, going from the gel band with the lowest molecular weight (C) to the one the next lowest weight (G) etc. The sequence read from the gel is complementary to the sequence in the template DNA and is indicated at the sites of the gel drawing.

template. Since this synthesis will not be synchronised, a set of oligonucleotides all starting from the primer but with a different length will be formed (Figure 16.2). After removal of the unreacted triphosphates, this oligonucleotide mixture, which is still hybridised to the template, is re-incubated with DNA polymerase in the absence of one of the four deoxyribonucleotides.

Consequently synthesis then proceeds until the missing deoxy-ribonucleotide should have been incorporated. Each chain therefore will terminate at its 3′ end at a position before the missing deoxy-ribonucleotide. Since separate samples are incubated, each with one of the four triphosphates missing, a complete set of oligonucletodes is obtained, ending before defined nucleotides and differing one nucleotide in length.

The newly synthesised strands are then separated from their template and subjected to electrophoresis in polyacrylamide gels in the presence of 8 M urea and the sequence can be read from the autoradiogram. The 'plus' system makes use of the fact that T4 DNA polymerase degrades double-stranded DNA from its 3′ end, and that this exonuclease activity is stopped at nucleotide residues that are present as their triphosphates in the incubation mixture. If this method is applied to the random oligonucleotide mixture as, obtained for the 'minus' method, all the chains will terminate on the nucleotide present in the incubation mixture during incubation with T4 DNA polymerase. After electrophoresis and autoradiography, the position of particular nucleotides will be indicated by bands one residue larger than in the corresponding 'minus' system. Since neither system is usually sufficient to establish a sequence, they are always used together. Probably because the DNA polymerase acts at different rates at different sites, expected products are frequently missing. A detailed protocol for the plus and minus method is given by Barrell.[5] The principal of the method is given in Figure 16.1.

The 'Dideoxy Method' for Chain Termination

More recently a method that uses specific chain terminating analogs of the normal deoxynucleoside triphosphates has been developed (Figure 16.2). The chain terminator method is considered to be the most simple, rapid and accurate method of its kind. It involves the synthesis of a complementary copy of the single-stranded DNA to be sequenced by the Klenow subfragment of DNA polymerase I.[6] This fragment lacks the 5′–3′ exonuclease activity of the intact enzyme. Priming is performed by annealing the directly adjacent strand of a restriction enzyme fragment.

Synthesis is carried out, in the presence of the four deoxyribonucleotide triphosphates, one of which is ^{32}P labelled. Chain termination is effected by the addition of 2′–3′-dideoxy or β-D-arabinofuranosyl analogues of the deoxyribonucleotide triphosphates to the reaction mixture. Dide-oxynucleotide triphosphates lack both 2′ and 3′ hydroxyl groups on the pentose ring. Since the DNA chain growth requires the addition of deoxynucleotides to the 3'-OH group, incorporation of a dideoxynucleotide terminates chain growth. Incorporation of the analogue in place of the normal substrate occurs randomly, so that the addition of a different dideoxynucleoside to one of four reaction mixtures generates a heterogeneous population of labelled strands terminating with the same nucleotide. The mechanism of chain termination is less obvious when an arabinose analogue is used. Arabinose is a steroisomer of ribose in which the 3' hydroxyl group is oriented in a *trans* position with respect to the 2' hydroxyl group. The arabinosyl nucleotides act as chain terminating inhibitors of *E. coli* DNA polymerase 1 in a way comparable to dideoxynucleotides. Chains ending in 3' ara nucleotides can however be further elongated by some mammalian DNA polymerases.[7]

Figure 16.2: The Chain Termination Method for Determination of the Nucleotide Sequence in DNA

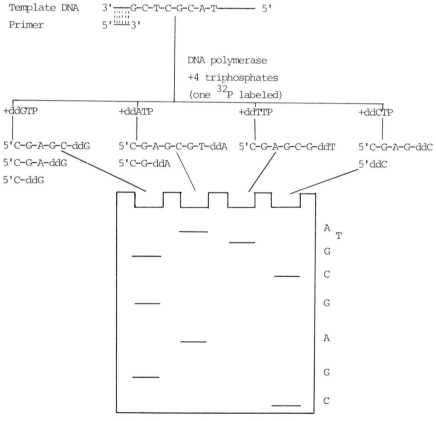

Autoradiogram of sequencing gel

Note: See note to Figure 16.1 for comments on how the sequence is read from the autoradiogram.

After termination of synthesis, the DNA from each of the four reaction mixtures is cleaved with a restriction endonuclease to separate labelled strands from their primer and is then denatured by heating with formamide. The single strands are separated electrophoretically in parallel on adjacent lanes of a gel, which is then autoradiographed. Since every possible fragment length will be represented in one of the four lanes the nucleotide sequence of the labelled strand can be read from the pattern of bands on the autoradiogram. The principle of the method is shown in Figure 16.2.

The chain terminating method has a number of advantages over the plus and minus method. It is more simple and rapid, since it only involves a single-step reaction. There is a better incorporation of counts from the ^{32}P-labelled triphosphates since the reaction can be carried out for long enough to allow extension from every primer. Finally and most importantly, every nucleotide shows up as a band, even in runs of the same nucleotide.

A requirement of the primed synthesis approach to DNA sequencing is a pure single-stranded template. This is a serious limitation for the general application of this sequencing technique since many double-stranded DNA molecules cannot be separated in pure single strands in good yield. Below the different stages of the primed synthesis method are discussed:

Preparation of Single-stranded Template DNA. A method for the preparation of pure single-stranded DNA has been described by Smith.[8] In this method linear double-stranded DNA molecules are digested by exonuclease III. This enzyme specifically digests double-stranded DNA from the 3' to the 5' end on both strands. The product of this digestion is two single-stranded DNA molecules, which have approximately half the number of nucleotides of the original double-stranded molecules. Since it is desirable to obtain sequences for the same DNA region from both strands, the fact that only single-stranded DNA is made from one of the two strands is a limitation to the use of exonuclease III in the generation of single-stranded templates. Other methods of preparing suitable template DNA are cloning of restriction enzyme fragments in a single-stranded bacteriophage vector,[9,10,11] alkaline cesium chloride ultracentrifugation, gel electrophoresis under denaturing conditions, or the poly (U G) method of Szybalki *et al*.

Annealing of Primer and Template and Primed Synthesis. A general protocol for the chain termination method for sequencing DNA, with dideoxynucleotides is given below.

A double-stranded restriction enzyme fragment, obtained from a preparative digest, is mixed with template in the appropriate buffer, in a ratio of primer to template of 3:7. The mixture is sealed in a capillary tube, heated at 100°C for 3 min in boiling water and incubated at 67°C for 30 min. After opening of the capillary, the annealed DNA solution is transfered to a siliconised test tube in which (α–^{32}P) deoxy ATP for labelling has been dried down under vacuum and the radioactive ATP is dissolved. The four dideoxy reaction mixtures are prepared in siliconised tubes. For details about the amount of deoxynucleotides and dideoxynucleotides to be used see Sanger *et al*. and Sanger and Coulson.[13,14] The priming reactions are also carried out in a capillary tube, in which a dideoxy mixture, the Klenow fragment of DNA polymerase, and annealed DNA are incubated at room temperature for 15 minutes.[15,16] (See the Instructions with the sequencing pack of New England Biolabs.)

Priming can be carried out using any size of restriction fragment. But with fragments less than 20 base pairs the concentration of primer has to be increased to get a similar incorporation of label. Since no carrier deoxy ATP is used in the dideoxy reaction mixtures (to keep the specific activity at the labelled dATP as high as possible) this can lead to chain termination due to lack of dATP. Therefore unlabelled dATP is added to each reaction and the prematurely terminated chains are then extended by incubation for another 15 minutes. The reactions are stopped by the addition of EDTA. The reaction mixtures, terminated by the addition of EDTA are then mixed with formamide dye mix, denatured at 100°C for three minutes and subjected to gel electrophoresis.

Gel Electrophoresis of the Four Sets of Labelled Oligonucleotides. Prior to gel electrophoresis, the primer used is either removed or remains bound to the newly synthesised DNA. When the restriction fragment used as a primer is large (100 nucleotides or more) it is removed by cleaving the DNA with the same restriction endonuclease whereby the fragment was generated. This can easily be done since the recognition site for the enzyme is restored during the primed synthesis. Large primers greatly increase the time needed for electrophoresis when bound to the newly synthesised radioactive DNA copies and reduce the resolution of the gel system. The preparation of the different types of sequence gels is described by Sanger and Coulson[17], Maat and Smith[18], Maxam and Gilbert[19] and Messing *et al*.[20] Normally there is enough sample to carry out several loadings on the same gel. After electrophoresis the gel is covered in Saranwrap and autoradiographed. Several percentages of polyacrylamide are routinely used for DNA sequence gels (i.e. 8%, 12% and 20% acrylamide) depending on the desired resolution, which is about 170 nucleotides for a 20% polyacrylamide slab gel, 40 cm in length and 1.5 mm thick and 250 nucleotides for a 10% gel of the same length but only 0.5 mm thick. The sets of oligonucleotides to be separated are denatured in a formamide containing dye mixture by heating at 100°C for three minutes. The dye mixture contains the bromphenol blue marker that corresponds to oligonucleotides of a chain length of 30 nucleotides. The xylene cyanol marker corresponds to a chain length of 80 nucleotides on an 8% gel. Electrophoresis is carried out at a constant power. The current ranges from 30–40 mA and the voltage from 1200–1700 volts depending on the thickness of the gel. Under these conditions the gels heat up to a temperature of 50–70°C, which is necessary to keep the DNA completely denatured, together with the 8 M urea that is also present in the gel. In order to prevent the glass plates in which the gel is held from breaking, due to these high temperatures, some laboratories use a thermostated gel electrophoresis sytem.

Two runs of the same set of oligonucleotides obtained by the dideoxy chain terminating method, run for different time periods on an 8% gel are shown in Figure 16.3.

The M13 Shotgun Strategy

As mentioned before, a pure single-stranded DNA template is essential if one wants to use the chain termination method of DNA sequencing. One of the methods to obtain such templates is the cloning of restriction enzyme fragments in single-stranded bacteriophage DNA.[21,22] In the earlier strategies, long DNA sequences were obtained by the use of a set of restriction fragment primers on a single template. The alternative of this strategy, the use of a set of recombinant templates with a single 'universal' primer is provided by the M13 shotgun strategy.[23]

Due to its special characteristics, the M13 phage is a very suitable cloning vehicle. M13 is a filamentous, male-specific coliphage that contains single-stranded circular DNA. Insertion of additional DNA into the phage genome results in a larger circular genome, which is then packaged in a longer filamentous protein coat. During infection the single-stranded DNA is converted to a double-stranded replicative form (RF), from which the single-stranded mature form is synthesised. Infection with M13 phage does not lead to lysis. The phage particles containing single-stranded DNA are continuously shed into the medium, but infected cells grow two to three times slower than uninfected cells. On agar plates the infected cells can therefore be seen as areas of retarded growth (plaques) in the lawn of cells. To facilitate cloning, and the screening for M13 phages containing foreign DNA, a region of the lactose operon of *E. coli* has been inserted into M13 RF DNA[24] and an *Eco* RI site was created in the gene coding for the α-peptide of β-galactosidase.

Using an *E. coli* strain as host, which has a defective chromosomal β-galactosidase gene, infection by M13 phage then allows complementation of the two β-galactosidase polypeptides and a functional β-galactosidase is produced. If grown on the appropriate indicator medium these cells will give rise to a blue plaque.[25] Infection with phages containing recombinant DNA, in which the foreign DNA is inserted at the unique *Eco* RI site within the α-peptide of β-galactosidase, gives rise to white plaques since the insertion of the foreign DNA destroys the possibility of complementation of the defective chromosomal β-galactosidase.

The recombinant phages are harvested by picking each white plaque from the plate and inoculating it in fresh medium. They are then grown up as small cultures. The single-stranded template required for sequencing DNA

Figure 16.3: Sequence Ladders Obtained with the Dideoxy Method for DNA Sequence Determination

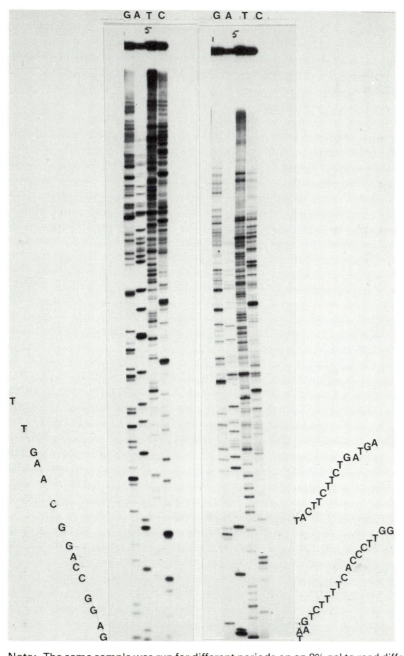

Note: The same sample was run for different periods on an 8% gel to read different regions of the sequence. Part of the sequence is given (Courtesy of Mary Duin van Raamsdonk and Rob Leer).

is easily prepared from an infected culture in less than two hours. The infected cells are spun down, the phage is precipitated from the supernatant with polyethylene glycol and the DNA is extracted with phenol. After precipitation with ethanol, usually enough template for sequencing is obtained from one ml of infected culture in this way.

Recently the M13 DNA has been further modified by introducing *Bam*HI, *Hind*II, *Sal*I, *Acc*I and *Pst*I sites into the β-galactosidase gene, allowing the direct cloning of DNA fragments generated by these enzymes. The successful insertion of such fragments is monitored also by the colour of the plaques.

The strategy for sequencing is then as follows. A large fragment of DNA is digested into many small fragments by single and/or multiple digestions with a restriction endonuclease. A digest is treated as a pool of fragments and cloned into M13 as a shotgun. Sequence data from independent clones, up to 250 nucleotides can be determined from the terminus of the fragment lying proximal to the priming site of the M13 vector, with the use of a universal primer. Several universal primers, complementary to the region of the M13 DNA at the 3'-end of the inserted DNA, are available today. The M13 shotgun strategy generates a random set of sequences from the region of interest. In the beginning most of these sequences will be different and data accumulation is very rapid; however at the end the accumulation of new sequences slows down dramatically. The sequences are usually entered into a computer and checked against each other for overlap. Several methods, that will not be discussed here, have been used to provide the required overlaps and to extend existing sequences. Today, probably the majority of DNA sequences are obtained with the M13 shotgun strategy described above.

3. The Maxam and Gilbert Method for DNA Sequencing

Introduction

The Maxam and Gilbert method for sequencing DNA is diagrammed in Figure 16.4. In this chemical procedure a (single-stranded) DNA fragment to be sequenced is end labelled (Figure 16.4A), usually by treatment with bacterial alkaline phosphatase to remove 5' phosphates, followed by reaction with ^{32}P-labelled ATP in the presence of polynucleotide kinase, which attaches ^{32}P to the 5' terminal nucleotide. Samples of labelled fragments are treated under four different conditions with chemical reagents that cause cleavage next to particular nucleotides. The four originally recommended conditions[26] are as follows

(i) Guanine cleavage. DNA is reacted with dimethylsulphate, which methylates the 7–position of guanine and the 3–position of adenine. Subsequent treatment with piperidine base breaks the phosphodiester bond next to positions that were occupied by guanine.

(ii) Guanine-adenine cleavage. DNA reacted with dimethylsulphate is treated with pyridinium formate at pH 2. Subsequent treatment of the acid-modified DNA with piperidine breaks the phosphodiester bond next to positions that were occupied by a purine. The dimethylsulphate reaction distinguishes adenines from guanines by the differences in their rates of methylation and depurination, gel bands resulting from guanine cleavage being stronger in the first and gel bands resulting from adenine cleavage being stronger in the second reaction.

(iii) Thymine-cytosine cleavage. Hydrazine reacts with DNA to cleave thymine and cytosine bases, forming ribosyl urea and other products called hydrazones. Treatment of the hydrazinolysis products with piperidine catalyses cleavage of the sugar phosphate backbone at exposed deoxyribose residues, thereby causing breakage of the DNA strand with about equal frequencies next to positions that were occupied by cytosine or thymine.

(iv) Cytosine cleavage. If the hydrazinolysis reaction is carried out in the presence of 5M sodium chloride, the reaction with thymine residues is preferentially suppressed. Cleavage with piperidine therefore specifically breaks DNA strands next to positions that were occupied by cytosines.

The conditions for these four reactions are chosen so that only a limited number of cleavages occur. This is achieved by the addition of unlabelled carrier DNA and variation of the reaction time. The extent to which bases should be modified in the sequencing reactions is defined by the resolution that can be obtained with a given gel system.

Ideally, reaction conditions should be such that only one base in each labelled molecule inside the region to be sequenced is modified. Since any base in the DNA for which a given reagent is specific is as likely to react as any other, these one-hit reactions evenly distribute the radioactivity among the cleavage products in the region one wants to sequence. The fragments produced in each reaction are separated as described before on the basis of chain length on polyacrylamide slab gels (Figure 16.4C), and the gel autoradiographed to indicate the positions of the radioactive bands (Figure 16.5). New versions of the four reactions, which are faster and provide salt-free cleavage products that can be used directly on the sequencing gels, have become available. A summary of these procedures together with some references where experimental details is given by Maxam and Gilbert.[27]

Practical Details

In our hands a satisfactory sequence was obtained with the chemical degradation method via the following protocol.

From the DNA region to be sequenced, a restriction enzyme fragment was prepared. The 5'-end phosphate group of this fragment was then exchanged for a ^{32}P-labelled phosphate group with the use of bacterial alkaline phosphatase, T4 polynucleotide kinase and ^{32}P ATP. Next the labelled fragment was cleaved with a second restriction enzyme that has to cleave the DNA at a position giving rise to two fragments with sufficient difference in length to be separated on a 5% polyacrylamide slab gel. Alternatively the DNA fragment generated by the first restriction enzyme was dissociated by melting the double-stranded DNA fragment under denaturing conditions and separated by gel electrophoresis into its complementary strands, which were then eluted (Figure 16.4A). Since double-stranded DNA with 5'-end single-stranded extensions is much more efficiently phosphorylated by T4 polynucleotide kinase than DNA with blunt or 3'-ends one of the restriction enzymes listed in Table 16.1 was normally used as the enzyme to generate the DNA fragment to be end labelled. Labelling of 3'-ends can be obtained with calf thymus terminal transferase and an (α–^{32}P) ribonucleotide triphosphate[28] or with DNA polymerase (either from T4 or *E. coli*) and (α–^{32}P) deoxynucleotide triphosphates.[29] The two fragments, labelled at only one end, obtained after incubation with the second restriction enzyme and electrophoresis, were eluted from the gel and sequenced. Labelling of the 5'-end of the restriction enzyme fragment was performed as follows. The DNA fragment (containing maximally 100 pmol terminal phosphate) was dissolved in 150 μl 50 mm Tris HCl buffer (pH 8.0) containing 1 mm EDTA. After addition of 5 μl of a solution of bacterial alkaline phosphatase the mixture was incubated for one hour at 65°C.

Prior to the addition of the phosphatase, the enzyme was freed of the ammonium sulphate present in the commercially available suspension. This was done either by pelleting the phosphatase from the ammonium sulphate suspension in an Eppendorf centrifuge and dissolving the pellet in Tris/HCl buffer (pH 8.0) or by dialysis. The incubation was performed in a 1.5 ml Eppendorf snapcap tube that was siliconised to prevent the DNA from sticking to the wall of the tube and sterilised to denature any DNAse that might be present in the tube. After the incubation the DNA was extracted several times with redistilled phenol, saturated with the incubation buffer and residual phenol was removed by repeated extractions with diethyl ether (three to four times).

The DNA was then precipitated. The DNA pellet was washed with 96% ethanol, dried under vacuo and dissolved in the incubation mixture for the polynucleotide kinase reaction. After addition of the γ-^{32}P ATP (50–150) μCi and the polynucleotide kinase (2 μl T4 polynucleotide kinase the mixture was incubated for 30–45 minutes at 37°C. The DNA was then precipitated three times to remove the excess labelled ATP. Prior to this 1 μl t-RNA (1 mg/ml) had been added to the DNA to stimulate co-precipitation. The DNA pellet was washed with 1 ml of 96% ethanol dried and dissolved in the buffer for the second restriction enzyme or in the sample buffer for strand separation. After incubation with the second restriction enzyme the DNA was precipitated and washed again and dissolved in 25 μl distilled water and 10 μl sample buffer.

Figure 16.4A: The Production of End Labelled DNA Fragments

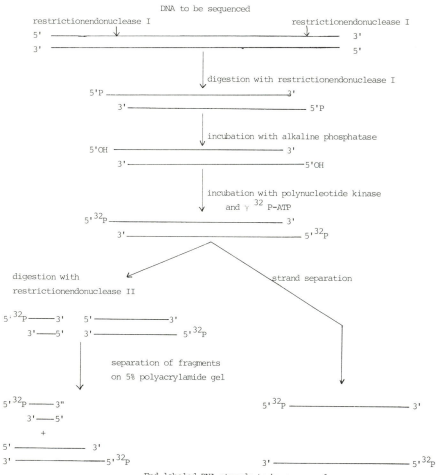

End labeled DNA strands to be sequenced

The labelled fragments were separated on a 5% polyacrylamide gel. The desired fragments were cut out, recovered from the gel by electro-elution or by soaking out the DNA for 15–20 h at 37°C. The eluate was filtered through siliconised glass-wool, plugged tightly in a 200 μl pipette tip, precipitated twice and dissolved in 20 μl distilled water. Five μl of this solution was used for each of the specific chemical cleavage reactions. After addition of 1 μg calf thymus DNA, modification of guanosine residues was performed with dimethylsulphate for 8–10 min. at 20°C, of adenosine residues with sodium hydroxide at 90°C for 10 min. and of cytosine and thymine residues with hydrazine for 10 min. at 20μC. Modification of cytosine alone was obtained with hydrazine in the presence of a high salt

Figure 16.4B: The Principle of Limited Modification of Nucleotides as Used in the Maxam and Gilbert Method

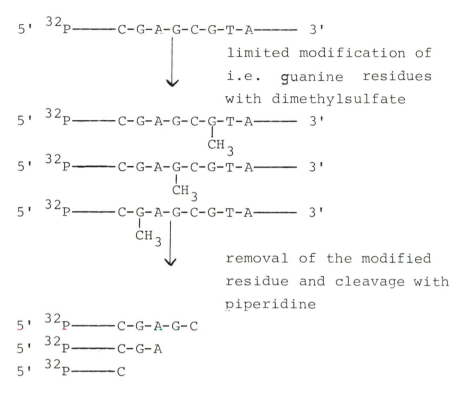

concentration. The reactions were stopped by the addition of specific stop buffers and the modified DNA was precipitated in ethanol (3 times). Strand scission was obtained with piperidine base by incubation for 30–45 min. at 90°C. The DNA was lyophilised to remove the piperidine and dissolved in water and lyophilised another three to four times. Finally the DNA pellet was dissolved in sample buffer and electrophorised as described before. Following electrophoresis autoradiography was performed at −80°C and the sequence read from the autoradiograph. In Figure 16.5 an example of such an autoradiograph is shown.

Discussion

In general the number of nucleotides that can be sequenced with the partial chemical degradation method and the chain termination method does not differ significantly. As said before, the latter method is probably the fastest of the two. However, if only a limited number of nucleotides is to be sequenced, or if one is just starting up DNA sequencing, the Maxam and

Figure 16.4C: Schematic Diagram of the Maxam and Gilbert Method for Sequencing

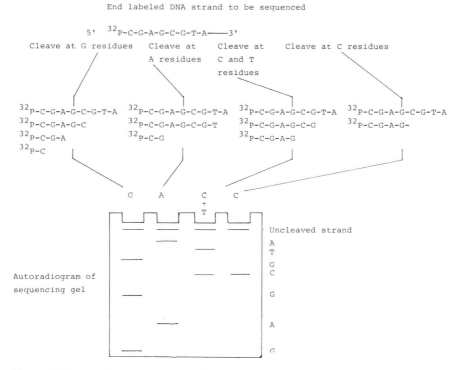

Note: In this case the sequence read from the autoradiogram is the actual sequence of the DNA fragment.

Gilbert method is to be preferred since the procedure is less complicated than the chain termination method. The amount of radioactivity incorporated by primed synthesis is normally much higher than by end labelling. This greatly reduces the time needed for exposing the film during autoradiography. One of the problems regularly encountered in the chain termination method is premature termination of synthesis at a region that the polymerase apparently finds difficult to copy. This gives rise to intense bands at the same position in all four tracks of the gel and makes it impossible to establish a sequence at that position.

One of the main problems encountered in the Maxam and Gilbert method are very weak thymine bands in the track for cytosine and thymine. Usually this is due to insufficient removal of salt from ethanol precipation. However, in both methods double bands sometimes arise, bands are found too close together, too far apart, too faint, too intense or missing. A trouble shooting list for the most common problems encountered is given in the first two of the suggested further readings.

The method for the determination of the nucleotide sequence of DNA have not only added many primary structures of proteins to the ones already determined by amino acid sequence methods, they also yielded some unexpected findings which could not have been discovered by amino acid sequence determination, namely the presence of so-called 'overlapping' genes, that is DNA regions that are coding for two genes.[30,31] It is not known yet whether overlapping genes are a general phenomenon or whether they are confined to viruses. Since with overlapping genes more genetic information is concentrated in a given-sized DNA, this phenomenon may be restricted to organisms whose survival depends on their rate of replication, and thus on the size of their genomes. Another unexpected finding, obtained when the first genes from vertebrates were sequenced, is the existence of spliced genes. The coding sequences for

Table 16.1: Restriction Endonucleases which Cleave DNA into Fragments with 5' Extensions, Readily End Labelled with T4 Polynucleotide Kinase

Enzyme	Sequence recognised
BamHI	5' GGATCC 3'
	3' CCTAGG 5'
BglII	5' AGATCT 3'
	3' TCTAGA 5'
BstEII	5' GGTNACC 3'
	3' CCANTGG 5'
EcoRI	5' GAATTC 3'
	3' CTTAAG 5'
HindIII	5' AAGCTT 3'
	3' TTCGAA 5'
HpaII	5' CCGG 3'
	3' GGCC 5'
HinfI	5' GANTC 3'
	3' CTNAG 5'
MboI	5' GATC 3'
	3' CTAG 5'
SalI	5' GTCGAC 3'
	3' CAGCTG 5'
Sau3AI	5' GATC 3'
	3' CTAG 5'
TaqI	5' TCGA 3'
	3' AGCT 5'
XbaI	5' TCTAGA 3'
	3' AGATCT 5'
XhoI	5' CTCGAG 3'
	3' GAGCTC 5'
XmaI	5' CCCGGG 3'
	3' GGGCCC 5'

Note: In this table N in a sequence stands for any of the four bases. The arrows indicate the points of cleavage.

Figure 16.5: Sequence Ladders Obtained with the Maxam and Gilbert Method for DNA Sequencing

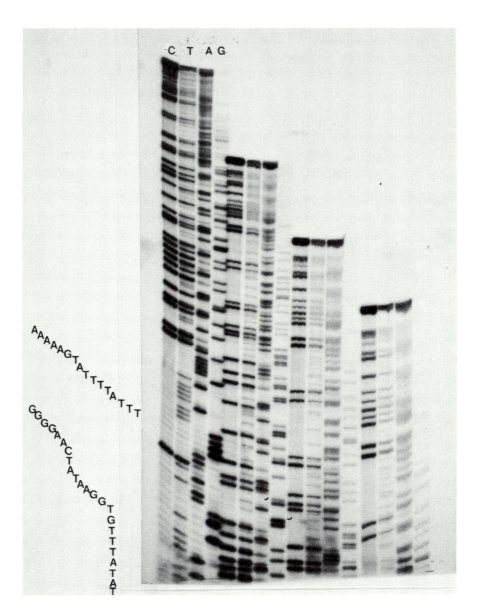

Note: The same sample was run on an 8% polyacrylamide gel different time periods. Part of the sequence that can be read from the autoradiogram is given.

globin, immunoglobin and ovalbumin do not lie on the DNA as a continuous series of codons, but they are interrupted by long regions of DNA, not coding for any protein. These regions are called introns and are much longer than the DNA regions leading to the expression of mature proteins, the exons. The nucleotide sequence in introns is rapidly changed by mutation, suggesting that their sequence is probably not relevant, but that their length, whereby they keep the exons apart along the chromosome, is their sole function.

There are several views concerning the evolution and the role of introns but it would lead too far to discuss these here. Although already known from amino acid sequence work, the synthesis of proteins in a precusor form containing a signalpeptide has been firmly established by the determination of the nucleotide sequence of the genes encoding them. This signal peptide is involved in the transport of the protein concerned across the cell membrane(s) and is cleaved off by special peptidases once the protein has reached its place of destination. Of course several other post-synthetic modifications of proteins are known, which cannot be detected by the DNA sequencing methods. In this respect, glycosidation, phosphorylation, ADP–ribosylation and modification of certain amino acid residues, i.e. methylation of lysine residues or hydroxylation of proline and serine residues, can be named. Until not too long ago, it was believed that the genetic code is universal. However, it was found by DNA sequencing that in human mitochondrial DNA the genetic code differs from the one used in *Escherichia coli* and yeast for example.[32] Comparison of a number of nucleotide sequences obtained for *E. coli* proteins revealed that in spite of the possibility of using six different codons for a particular amino acid (the degeneracy of the genetic code) the organism has a strong preference for only one or two of the possible codons.[33] The same phenomenon has been shown for codon usage in yeast.[34]

Both amino acid sequencing and DNA sequencing are methods that are rather expensive. For protein sequencing, specially purified chemicals are needed, whereas for DNA sequencing ^{32}P-labelled nucleotides are needed. In general it can be said that DNA sequencing is a lot faster than protein sequencing, especially if one takes into account the laborious work of protein and peptide purification needed to obtain a complete protein sequence, even though the latter has been automated. However this is only taking into account the actual sequencing and isolation of the DNA fragments. When nothing about the location of the gene for the protein that one wants to sequence is known, its location may be quite time consuming. An experienced worker can sequence six DNA fragments in two weeks and 200–230 nucleotides per fragment can be obtained to give a rough indication of the amount of time involved. However, the need for some amino acid sequence data, i.e. N- and C-terminal sequences, will exist, since without these the correct reading frames of the DNA sequences, the primary structure of a protein, is hard to establish.

Further Reading

1. A.M. Maxam and W. Gilbert, 'Sequencing End Labelled DNA with Base Specific Chemical Cleavages in L. Grossman and K. Moldave (eds.), *Methods in Enzymology* (Academic Press, New York, 1980), 65, 499–560.
2. A.J.H. Smith, 'DNA Sequence Analysis by Primed Synthesis', in L. Grossman and K. Moldave (eds), *Methods in Enzymology*, (Academic Press, New York, 1980), 65, 560–80.
3. F. Sanger, 'Determination of Nucleotide Sequences in DNA', *Science*, *214* (1981), 1205–10
4. W. Gilbert, 'DNA sequencing and Gene Structure', *Science*, *214* (1981), 1305–12.

References

1. F. Sanger and A.R. Coulson, 'A rapid method for determining sequences in DNA by primed synthesis with DNA polymerase', *J. Mol. Biol.*, *94* (1975), 441–8.
2. F. Sanger, S. Nicklen and A.R. Coulson, 'DNA sequencing with chain-terminating inhibitors', *Proc. Natl. Acad. Sci. USA*, *74* (1977), 5463–7.
3. A.M. Maxam and W. Gilbert, 'A new method for sequencing DNA', *Proc. Natl. Acad. Sci. USA*, *74* (1977), 560–4.
4. Sanger and Coulson, 'A rapid method for determining sequences'.
5. B.G. Barrell, 'Biochemistry of Nucleic Acids' in B.F.C. Clark (ed.), *Int. Rev. Biochem.*, (University Park Press, Baltimore, 1978), vol. 17, 125–48.
6. H. Klenow, K. Overgaard-Hanesen and S.A. Patkar, 'Proteolytic cleavage of native DNA polymerase into two different catalytic fragments', *Eur. J. Biochem.*, *22* (1971), 371–81.
7. T. Hunter and B. Francke, 'In vitro polyoma DNA synthesis: Inhibition by a α–β–D arabino faranosyl CTP', *J. Virol.*, *15* (1975), 759–75.
8. A.J.H. Smith, 'The use of Exonuclease III for preparing single-stranded DNA for use as a template in the chain terminator sequencing method', *Nucleic Acids. Res.*, *6* (1979), 831–48.
9. F. Sanger, A.R. Coulson, B.G. Barrell, A.J.H. Smith and B.A. Roe, 'Cloning in single-stranded bacteriophage as an aid to rapid DNA sequencing', *J. Mol. Biol.*, *143* (1980), 161–78.
10. J. Messing, B. Gronenborn, B. Muller-Hill and P.H. Hofschneider, 'Filamentous coliphage M13 as a cloning vehicle: Insertion of a Hind II fragment of the *lac* regulatory region in M13 replicative form *in vitro*', *Proc. Natl. Acad. Sci. USA*, *74* (1977), 3642–3646.
11. W. Szybalki, H. Kubinski, Z. Hradecna and W.C. Summers, 'Analytical and Preparation separation of the complementary DNA strands' in L. Grossman and K. Moldave (eds.), *Methods in Enzymology* (Academic Press, New York, 1971), 21D, 382–413.
12. A.J.H. Smith, 'DNA sequence analysis by primed synthesis' in L. Grossman and K. Moldave (eds.), *Methods in Enzymology* (Academic Press, New York, 1980), 65, 560–80.
13. Sanger, Nicklen and Coulson, 'DNA sequencing'.
14. F. Sanger and A.R. Coulson, 'The use of thin acrylamide gels for DNA sequencing', *FEBS Letts*, *87* (1978), 107–10.
15. Sanger, Nicklen and Coulson, 'DNA sequencing'.
16. Sanger and Coulson, 'The use of thin acrylamide gels'.
17. Ibid.
18. J. Maat and A.J.H. Smith, 'A method for sequencing restriction fragments with dideoxy nucleoside triphosphates', *Nucleic Acids Res.*, *5* (1978), 4537–46.
19. A.M. Maxam and W. Gilbert, 'Sequencing end labelled DNA with base specific chemical cleavages' in L. Grossman and K. Moldave (eds.), *Methods in Enzymology* (Academic Press, New York, 1980), vol. 65, 499–560.
20. J. Messing, R. Crea and P.H. Seeburg, 'A system for shotgun DNA sequencing', *Nucleic Acids Res.*, *9* (1981), 309–20
21. Sanger, Coulson, Barrell, Smith and Roe, 'Cloning'.
22. Messing, Gronenborn, Muller-Hill and Hofschneider, 'Filamentous coliphage M13'.
23. Szybalki, Kubinski, Hradecna and Summers, 'Analytical and Preparative sequencing'.

24. R. Roychoudbury and R. Wu, 'Terminal transferase catalyzed addition of nucleotides to the 3' termini of DNA' in L. Grossman and K. Moldave (eds.), *Methods in Enzymology* (Academic Press, New York, 1980), 65, 43–62.
25. Messing, Gronenborn, Muller-Hill and Hofschneider, 'Filamentous coliphage M13'.
26. Roychoudbury and Wu, 'Terminal transferase'.
27. Maxam and Gilbert, 'A new method for sequencing DNA'.
28. Ibid.
29. M.D. Challberg and P.T. Englund, 'Specific labelling of 3' termini with T4 DNA polymerase', *Methods in Enzymology* (Academic Press, New York, 1980), 65, 39–43.
30. F. Sanger, G.M. Air, B.G. Barrell, N.L. Brown, A.R. Coulson, J.C. Fiddes, C.A. Hutchison, P.M. Slocombe and M. Smith, 'Nucleotide sequence of bacteriophage φ X174 DNA', *Nature*, 265, 687–95.
31. F. Sanger, A.R. Coulson, T. Friedmann, G.M. Air, B.G. Barrell, N.L. Brown, J.C. Fiddes, C.A. Hutchison, P.M. Slocombe and M. Smith, 'The nucleotide sequence of bacteriophage Ø X174', *J. Mol. Biol.*, 125 (1978), 225–46.
32. B.G. Barrell, A.T. Bankier and J. Drouin, 'A different genetic code in human mitochondria', *Nature*, 282 (1979), 189–94.
33. R. Grantham, C. Gautier, M. Gouy, R. Mercier and A. Pavé, 'Codon catalog usage and the genome hypothesis', *Nucleic Acids Res.*, 8 (1980), r49–r62.
34. J.L. Bennetzen and B.D. Hall, 'Coden selection in Yeast', *J. Biol. Chem.*, 257 (1981), 3026–3031.

17 THE USE OF COSMIDS AS CLONING VEHICLES

Jan van Embden

CONTENTS

1. Introduction

The use of plasmid vectors in the cloning of very large segments of DNA is hampered by the low transformation efficiency of large plasmids. Bacteriophage lambda can be used as a vector to clone efficiently larger segments of DNA. Lambda vectors can accept at most 18–21 kilobases (kb) of foreign DNA.[1] More recently Collins and Hohn[2] have developed a new type of plasmid vector which combines the high efficiency of transfection with packaged 'phage' particles and the advantage of using a plasmid vector to accept larger segments of foreign DNA. This new type of vector differs from the usual plasmid vectors in the presence of a small piece of lambda DNA, the so-called cohesive end site. Lambda DNA in the virus is linear with two single-stranded ends of twelve nucleotides in length. Because these ends are complementary, they are called the cohesive ends or *cos* sites. Such vectors are, therefore, designated as cosmids. Cosmid hybrids can be packaged into lambda phage heads and this allows the cloning of the DNA fragments up to 52 kb. After transduction into a lambda-sensitive bacterium the hybrid plasmids replicate as plasmids and are selected for by using antibiotic resistance markers mediated by these vectors. The efficiency by cosmid cloning can be as high as 10^6 clones per μg of foreign DNA. This is comparable to the efficiencies obtained with lambda vectors. An important advantage of cosmid cloning is that the system selects specifically the cloning of large DNA fragments, in contrast to transformation, which tends to favour the cloning of smaller DNA fragments.

2. Lambda Morphogenesis and DNA Requirements for Packaging

Studies on the morphogenetic pathway of phage lambda (for a review see Hohn and Katsura[3]) made it possible to package DNA *in vitro* by supplying preheads, certain phage-specific DNA-binding proteins and a polyamine. The earliest precursor in the head morphogenesis of lambda is the scaffolded prohead which is composed of about 400 copies of protein E and three scaffolding proteins. The scaffolding proteins are removed with the help of a host protein resulting in the mature prehead, which is about 20% smaller than the phage head. The mature prohead can recognise selectively the phage DNA, due to the binding of the lambda proteins A and Nu1 to the lambda DNA. The DNA recognition site for these proteins is a sequence close to the left cohesive end of lambda. The packaging of the DNA is accompanied by an enlargement of the prohead and cleavage of the precursor DNA at the *cos* sites, generating the cohesive end termini of the packaged DNA molecule. This so-called terminase (Ter) reaction results in the introduction of two nicks staggered twelve nucleotides apart on the opposite strands of the duplex. Subsequently, the lambda protein, D, aggregates to the capside and the independently assembled tail is attached to the mature phage head. This results in infective particles that can inject the DNA into bacterial cells. The injected DNA circularises in the cell due to the presence of the cohesive ends.

The DNA substrate for *in vivo* packaging of lambda DNA consists of tandem polymers that result from replication. Multimeric as well as monomeric linear lambda DNA is packageable *in vitro*. Circular DNA can be packaged when at least two *cos* sites are present on the molecule.[4]

The systematic study of the DNA specific requirements for packaging of DNA learned that any DNA can be packaged *in vitro* when it fulfills the following requirements: (i) The presence of a DNA sequence near the *cos* site, which might be recognised by the proteins A and Nu1; (ii) The presence of at least 2 *cos* sites, which are the sequences recognised by the protein A-containing terminase; (iii) The distance between 2 successive *cos* sites should be in the range of 37 to 52 kilobases, because it is only in this range that a high efficiency of packaging to infectious particles is obtained.

Collins *et al.*[5,6,7,8] exploited the knowledge on these specific requirements to develop plasmid vectors that carry the *cos* region. These circular plasmids themselves cannot be packaged; however, after linearisation by cleavage with restriction enzymes and ligation to foreign DNA fragments, packageable hybrids can be formed consisting of foreign DNA flanked by vector DNA. Because of the size requirement for packaging, the DNA fragments to be cloned will selectively be in the range of 37 to 52 kb minus the size of the vector. The conditions during ligation are critical in any cosmid cloning experiment, because concatemers consisting of vector DNA or mainly vector DNA will also be packaged with high efficiency. Therefore, these conditions should be chosen so as to minimise the resulting background of transductants carrying only vector DNA or only small segments of DNA to be cloned.[9,10] A schematic view of a typical cosmid cloning experiment is shown in Figure 17.1.

3. The Packaging System

Various procedures have been described to package DNA into lambda phage heads. All are based on mixtures of complementary cell extracts obtained from different induced lambda lysogens, each of which is blocked in a differenct gene for phage head morphogenesis.

The simplest procedure described is that of Collins and Hohn.[11] They use thermoinducible phages having amber mutations in the structural phage

Figure 17.1: A Typical Cosmid Cloning Experiment

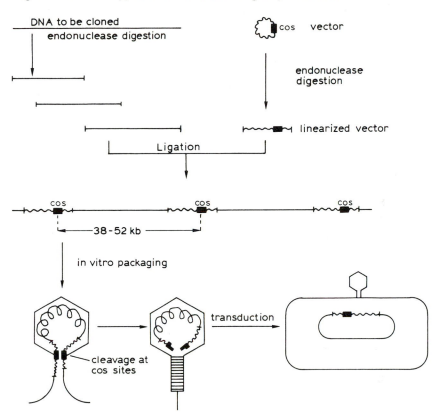

Note: Target DNA to be cloned is cleaved by restriction endonuclease digestion to result in fragments of about 40 kb. These are ligated to linearised cosmid DNA and this mixture is *in vitro* packaged. Only molecules having two *cos* sites at a distance of 37 to 52 kb will be packaged into the phage heads. As a consequence of packaging the DNA is cleaved by the *ter* protein, to give cohesive sites, that enable the recombinant cosmid molecule to recirculise after injection into a host cell.

head proteins E and D, respectively. Both phages carry, in addition, a mutation in the lysis gene S, so that cultures fail to lyse spontaneously and phage proteins accummulate several hours after heat induction. To diminish the amount of packageable lambda DNA in heat induced extracts, the phages carry the b2 deletion that prevents lambda DNA from excising from the chromosome. Finally, the mutation in the lambda recombination gene, *red*3, and the use of a recombination-deficient (*rec*A) host strain minimise recombination that might result in the production of phages carrying markers present on both endogenous and exogenous DNA.

Cells of the two heat-induced cultures are mixed and concentrated in a buffer containing spermidine, putrescine, beta-mercaptoethanol, dimethyl sulphoxide and ATP. The mixture is frozen in liquid nitrogen and the packaging mix can be stored at −56°C for long periods of time without significant loss of activity. For packaging, the mix is placed on ice for a few minutes and mixed with the DNA to be packaged. By raising the temperature to 25°C or 37°C the cells lyse and liberate all components needed for packaging. After 20–60 minutes of incubation the mixture is diluted and the remaining viable cells are killed by adding a drop of chloroform. Transduction is done with recipients that are grown in the presence of maltose in order to induce the lambda receptor protein. Optionally, induced cultures are irradiated by ultraviolet light to destroy packageable lambda DNA in the packaging mix.[12,13] A problem in this procedure is the high viscosity of the packaging extract that hampers appropriate mixing with the DNA to be packaged. A convenient way to overcome this problem is the procedure of V.Pirotta (personal communication) who simplified the lambda packaging method of Blattner.[14] In this procedure the induced E⁻ lambda lysogen is used to make a freeze-thaw lysate (FTL) that is centrifuged at 30,000 r.p.m. to pellet the DNA. A concentrated induced culture of the D⁻ lambda lysogen is sonicated to destroy the DNA. It also destroys some of the phage tails and phage fibres. This sonicated extract (SE) will, therefore, lack tail components and protein D. The FTL provides intact phage tails as well as protein D, but in turn is missing protein E. Both extracts are stored separately at −70°C. For packaging, the optimal ratio of FTL and SE has to be determined for each preparation. The details of Pirotta's procedure are described by Grosveld *et al.*[15]

4. Cosmid Vectors

Cosmid vectors can easily be constructed by insertion of the *cos* region of lambda into existing cloning vehicles. The size of this lambda sequence can be as small as 400 base pairs.[16] The first series of cosmids as cloning vehicles were constructed by Collins and co-workers [17,18] and these were six to 24 kb in size. These vectors allow the cloning of DNA having sizes ranging from 28 to 46 kb, depending on the choice of the vector. Later they constructed cosmid pHC79 which is a *cos*-containing derivative of the frequently used vector pBR322.[19] Somewhat smaller cosmids were obtained by Meyerowitz *et al.*[20] who inserted a small lambda *cos* segment of approximately 400 base pairs at the PstI site of the versatile cloning vehicle pBR322.

Chia *et al.*[21] constructed a series of cosmid vectors, called Homer cosmids, derived from pAT153, a safe derivative of pBR322 favoured for reasons of biological containment. One of them, Homer III, contains sequences required for the replication and transcription of DNA of the eukaryotic virus SV40 and it lacks pBR322 sequences which inhibit efficient replication in eukaryotic cells.

Ish-Horowicz and Burke[22] constructed the cosmid pJB8, in which a synthetic *Bam*HI site has been inserted into the *Eco*RI site of Homer I (see Table 17.1). The plasmid allows the cloning into the *Bam*HI site of DNA fragments produced by partial digestion with *Mbo*I or *Sau*3A and the inserted DNA can be excised from hybrid cosmids by *Eco*RI digestion.

The above described cosmids are derived from multicopy plasmids. Certain DNA segments are difficult to clone on high copy number vectors, because they encode for products which are not tolerated at high cellular levels in host bacteria.[46] In such cases it might be recommended to use low copy number vectors like pHSG422, which carries the replicon of plasmid pSC101.[47]

A number of versatile cosmid vectors and their properties are listed in Table 17.1.

5. The Conditions During Ligation

The purpose of cloning experiments with cosmid vectors is usually the cloning of large contiguous chromosomal DNA fragments. Therefore, the conditions during ligation should be chosen so as to minimise the self-ligation of chromosomal DNA fragments, that could result in clones carrying non-contiguous chromosomal DNA fragments. This can be achieved by taking care not to have chromosomal DNA fragments that are smaller than the desired size range to be cloned.[48] Best results are obtained by fractionation of partially digested DNA (for example by centrifugation through density gradients), although merely partial digestion might yield satisfactory results as well. If high cloning efficiencies are required, the target DNA fragments can be dephosphorylated to avoid any self-ligation. A high concentration of DNA to be cloned and an excess of vector DNA, in the range of 300 to 1000 μg/ml, favour the efficient formation of long concatameric molecules of target DNA intersperced by *cos* sequences.[49,50,51]

However, also self-ligation of vector molecules might result in concatemeric vector oligomers of packageable size and/or in packageable molecules consisting of target DNA, each end of which is linked to vector DNA. Both kinds of molecules might generate cosmid monomers due to intramolecular recombination in the host cell even when it is recombination deficient.[52]

Ish-Horowicz and Burke[53] have designed an elegant method to avoid chromosome scrambling in cosmid hybrids and to rule out the insertion of smaller chromosomal segments due to vector to vector ligation. They used pJB8 to make a left-hand vector *cos* fragment that cannot ligate in tandem but which accepts dephosphorylated target DNA. This was done by cleavage of pJB8 with *Hind*III, dephosphorylation and cutting with *Bam*HI (see Figure 17.2). Similarly, a right-hand vector *cos* fragments was produced by cleavage with *Sal*I, dephosphorylation and subsequent cleavage with

Table 17.1: Properties of some Cosmid Vectors

Cosmid	Size (kb)	Resistance[a]	Unique sites, suitable for insertion of target DNA	Gene inactivation	Reference[e]
pJC720	24.3	Rif	Xmal, HindIII	Rif	23
pJC74	16	Ap	EcoRI, BamHI, BglII	–	24,25
pJC75.58	11.5	Ap	EcoRI, BamHI, BglII	–	26,27
pHC79[c]	6.4	Ap,Tc	EcoRI, ClaI, AvaI, HpaI, EcaI, BalI, PvuII		28,29,30,31,32,33
			HindIII, BamHI, SalI	Tc	
			PstI, PvaI	Ap	
pJB8[c]	5.5	Ap	same as pHC79[b]	–	34,35,36
pMUA3[c]	4.9	Tc	EcoRI, ClaI, AvaI, BalI, PvuI	–	37,38
			HindIII, BamHI, SalI		
pTL5[c]	5.8	Tc	ClaI, BglII, HpaI, EcaI	–	39
			HindIII, BamHI, SalI		
Homer I[c]	5.4	Ap	HindIII, SalI, ClaI, EcoRI, PstI, PvuI	–	40,41,42
Homer II[c]	6.4	Ap	SalI, EcoRI, SstI	–	43
pFF2	11.1	Km	EcoRI, PstI, BamHI, SalI	–	44
pHSG422[d]	8.8	Cm,Km,Ap	BstEII		45
			PstI	Ap	
			HindIII, XmaI, XhoI	Km	
			EcoRI	Cm	

Notes: a. Rif = rifampicin; Ap = ampicillin; Tc = tetracycline; Cm = chloramphenicol; Km = kanamycin
b. pJB8 carries in fact two close EcoRI sites that flank the BamHI site. This permits EcoRI excision of Sau3A or MboI fragments cloned in the BamHI site.
c. These cosmids are derivatives of the cloning vehicle pBR322.
d. Low copy number, thermosensitive replication.
e. The first reference refers to the construction of the cosmid vector.

Figure 17.2: Efficient Cosmid Cloning According to Ish-Horowitz and Burke.[54]

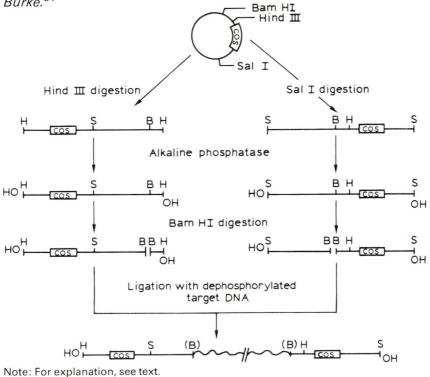

Note: For explanation, see text.

*Bam*HI. A mixture of left-hand fragments and right-hand fragments cannot result in packageable molecules, because only molecules with 2 *cos* sites separated by 37–52 kb can be packaged. The target DNA is dephosphorylated to prevent the self-ligation of target DNA molecules. A mixture of equimolecular left- and right-ended vector fragments and dephosphorylated target DNA thus gives only one kind of packageable molecules. Because of the size-specificity of the packaging reaction, no fractionation to correct size of target DNA fragments is needed. This rapid and efficient cosmid cloning method can be used for any cosmid vector, when merely one of the sites used to make the vector arms is located between the *cos* site and the insertion site of DNA to be cloned.

6. Applications

The main advantage in the use of cosmids for the molecular cloning of DNA is the efficient cloning of high molecular DNA. Therefore, the construction of gene libraries is most efficiently done by cosmid cloning. To obtain a gene bank containing once the entire genome of *Escherichia coli, Saccharomyces cerevisiae, Drosophila* or mouse would require 1.2×10^2, 4×10^2, 4×10^3 and 7×10^4 clones, respectively, which contain on average a 40 kb insert.[55] To avoid the exclusion of very large or very small DNA fragments that are produced by restriction enzymes with six base recognition sites, it is recommended to use partial digests with multiply cutting enzymes.[56] These will produce overlapping fragments of the entire genome.

The cosmid system is particularly useful to clone flanking sequences if a DNA segment that has already been cloned, and which can be used as a probe for hybridisation. By repeating this process using the newly cloned DNA as a probe, it is possible to 'walk' along long stretches of cellular DNA. Lund *et al.*[57] used cosmids to rescue large segments of transforming yeast DNA that contained a 'transforming gene' linked to the bacterial beta-lactamase gene that can be selected for during the cosmid cloning procedure.

The efficiency of cosmid cloning is high and it is comparable to that by cloning with lambda vectors. The various authors report efficiencies of 10^3 to 10^6 transducing particles per µg target DNA. This latter value is several orders of magnitude higher than the efficiency of transformation of plasmids of 40 kb. The method may be particularly useful in combination with methods that allow the *in situ* screening of plates, with high colony densities like that developed by Hanahan *et al.*[58] It should be emphasised that the plasmid copy number in hybrid cosmid clones is usually lower than that of the vector and, therefore, they will produce a lower radioactive signal using *in situ* filter hybridisations. In certain cases better results may be obtained at a not too high colony density.[59]

Most authors observed recombinant cosmid DNA in the clones smaller than the size that lambda heads will normally package. These might have arisen by intramolecular recombination during growth, even in *rec*A cells.[60] Although cosmids are usually stable in Rec$^+$ as well as in *rec*A hosts, occasionally instability is observed in a Rec$^+$ host, whereas the same plasmid is perfectly stable in a *rec*A host.[61] The use of a *rec*A host might result in larger mean inserts than in Rec$^+$ hosts.[62] Therefore, *rec*A hosts are particularly recommended for cosmid cloning experiments.

The DNA recovery of hybrid DNA from cosmid clones can vary greatly. The various authors report yields of 0.1 to 2 µg of DNA per ml of culture. In our laboratory we observed occasionally a much lower yield of DNA, due to extreme instability of the hybrids. When plasmids delete during subsequent cycles of growth, one might recover the original cosmid hybrid by repacking the original recombinant DNA preparation, followed by transduction.[63]

The variety of cosmids listed in Table 17.1 is rather limited. It should be kept in mind that any *E. coli* plasmid vector is easily transformed into a cosmid vector just by insertion of the small *cos* region into that vector; so cosmid vectors can rapidly be developed to meet specific cloning requirements. The host range of cosmids has been extended to *Salmonella typhimurium*, which normally does not have the lambda receptor. This was done by introduction of the *malB* region of *E. coli* K12 into *S. typhimurium*.[64] *MalB* is the gene that encodes for the lambda receptor protein. It is conceivable to further extend the host range of packaged cosmid hybrids to other hosts in a similar way.

An elegant application of the use of cosmids is the method of Lindenmayer *et al.*[65] to move cloned DNA into and out of eukaryotic cells. In this method 'gene shuttling' the ability of the lambda terminase to cut DNA in the *cos* sites is exploited at various steps. To cleave recombinant cosmid molecules from a gene library at unique sites, *in vivo* packaging of these molecules was done. The presence of two *cos* sites in tandem permitted *in vivo* packaging of recombinant cosmid DNA. Transformation of mouse cells by this linearised DNA led to the establishment and expression of the recombinant cosmid DNA. To reclone this DNA from the mouse cells the use of restriction enzymes could be avoided. Just adding uncleaved total mouse cell DNA as a substrate for the *in vitro* packing system resulted in recutting and packaging of the DNA to be recloned.

Further Reading

1. J. Collins, '*Escherichia coli* plasmids packageable *in vitro* in λ bacteriophage particles', *Methods in Enzymology* (Academic Press. New York 1979), 68, 309–26.
2. J. Collins and H.J. Brüning, 'Plasmids useable as gene-cloning vectors in an in vitro packaging by coliphage λ: "cosmids"', *Gene*, *4* (1978), 85–107.
3. T. Hohn and I. Katsura, 'Structure and assembly of bacteriophage lambda', *Curr. Top. Microbiol. Immunol.*, *78* (1977), 69–110.
4. D. Ish-Horowicz and J.F. Burke, 'Rapid and efficient cosmid cloning', *Nucleic Acids Res.*, *9* (1981), 2989–98.
5. F.G. Grosveld, H.-H.M. Dahl, E. de Boer and R.A. Flavell, 'Isolation of β-globin-related genes from a human cosmid library', *Gene*, *13* (1981), 227–37.

References

1. F.R. Blattner, A.E. Blechl, K. Denniston-Thompson, H.E. Faber, J.E. Richards, J.L. Slightom, P.W. Tucker and O. Smithies, 'Cloning human fetal γ globin and mouse α-type globin DNA: preparation and screening of shotgun collections', *Science*, *202* (1978), 1279–84.
2. J. Collins and B. Hohn, 'Cosmids: a type of plasmid gene-cloning vector that is packageable *in vitro* in bacteriophage λ heads', *Proc. Natl. Acad. Sci. USA*, *75* (1978), 4242–6.
3. T. Hohn and I. Katsura, 'Structure and assembly of bacteriophage lambda', *Curr. Top. Microbiol. Immunol.*, *78* (1977), 69–110.
4. Ibid.
5. Collins and Hohn, 'Cosmids'.
6. J. Collings and H.J. Brüning, 'Plasmids useable as gene-cloning vectors in an *in vitro* packaging by coliphage λ: "cosmids"', *Gene*, *4* (1978), 85–107.
7. B. Hohn and J. Collins, 'A small cosmid for efficient cloning of large DNA fragments', *Gene*, *11* (1980), 291–8.
8. J. Collins, '*Escherichia coli* plasmids packageable *in vitro* in λ bacteriophage particles' in

S.P. Colowick and N.O. Kamplan (eds.), *Methods in Enzymology, Recombinant DNA* (Academic Press, New York, 1979), vol. 68, 309–26.
9. Collins and Brüning, 'Plasmids useable as gene-cloning vectors'.
10. Collins, '*E.coli* plasmids'.
11. Collins and Hohn, 'Cosmids'.
12. Collins and Brüning, 'Plasmids useable as gene-cloning vectors'.
13. Collins, '*E.coli* plasmids'.
14. Blattner, Blechl, Denniston-Thompson, Faber, Richards, Slightom, Tucker and Smithies, 'Cloning human fetal γ globin'.
15. F.G. Grosveld, H.-H.M. Dahl, E. de Boer and R.A. Flavell, 'Isolation of β-globin-related genes from a human cosmid library', *Gene, 13* (1981), 227–37.
16. E.M. Meyerowitz, G.M. Guild, L.S. Prestidge and D.S. Hogness, 'A new high-capacity cosmid vector and its use', *Gene, 11* (1980), 271–82.
17. Collins and Hohn, 'Cosmids'.
18. Collins and Brüning, 'Plasmids useable as gene-cloning vectors'.
19. Hohn and Collins, 'A small cosmid for efficient cloning'.
20. Meyerowitz, Guild, Prestidge and Hogness, 'A new high-capacity cosmid vector'.
21. W. Chia, M.R.D. Scott and P.W.J. Rigby, 'The construction of cosmid libraries of eukaryotic DNA using the Homer series of vectors', *Nucleic Acids Res.*, *10* (1982), 2503–20.
22. D. Ish-Horowicz and J.F. Burke, 'Rapid and efficient cosmid cloning', *Nucleic Acids Res.*, *9* (1981), 2989–98.
23. Collins and Brüning, 'Plasmids useable as gene-cloning vectors'.
24. Ibid.
25. R. Ollo, C. Auffray, C. Morchamps and F. Rougeon, 'Comparison of mouse immunoglobulin gamma2a and gamma2b chain genes suggests that exons can be exchanged between genes in a multigenic family', *Proc. Natl. Acad. Sci. USA*, *78* (1981) 2442–6.
26. Collins and Brüning, 'Plasmids useable as gene-cloning vectors'.
27. F. Gannon, J.M. Jeltsch and F. Perrin, 'A detailed comparison of the 5'-end of the ovalbumin gene cloned from chicken oviduct and erythrocyte DNA', *Nucleic Acids Res.*, *8* (1980), 4405–21.
28. Hohn and Collins, 'A small cosmid for efficient cloning'.
29. J.R. Arrand, L. Rymo, J.E. Walsh, E. Björck, T. Lindahl and B.E. Griffin, 'Molecular cloning of the complete Epstein-Barr virus genome as a set of overlapping restriction endonuclease fragments', *Nucleic Acids Res.*, *9* (1981), 2999–3014.
30. R. Cattaneo, J. Gorski and B. Mach, 'Cloning of multiple copies of immunoglobulin variable kappa genes in cosmid vectors', *Nucleic Acids Res.*, *9* (1981), 2777–90.
31. E. May, J.-M. Jeltsch and F. Gannon, 'Characterisation of a gene encoding a 115 K super T antigen expressed by a SV40-transformed rat cell line', *Nucleic Acids Res.*, *9* (1981) 4111–27.
32. E.T. Palva and P. Liljeström, 'Cosmid cloning and transposon mutagenesis in *Salmonella typhimurium* using phage λ vehicles', *Mol. Gen. Genet.*, *181* (1981), 153–7.
33. M. Schweizer, M.E. Case, C.C. Dijkstra, N.H. Giles and S.R. Kushner, 'Identification and characterisation of recombinant plasmids carrying the complete *qa* gene cluster from *Neurospora crassa* including the *qa-1* regulatory gene', *Proc. Natl. Acad. Sci. USA*, *78* (1981), 5086–90.
34. Ish-Horowicz and Burke, 'Rapid and efficient cosmid cloning'.
35. G. Gross, U. Mayr, W. Bruns, F. Grosveld, H.-H.M. Dahl and J. Collins, 'The structure of a thirty-six kilobase region of the human chromosome including the fibroblast interferon gene IFN-β', *Nucleic Acids Res.*, *9* (1981), 2495–2507.
36. E.H. Weiss, K.S.E. Cheah, F.G. Grosveld, H.-H.M. Dahl, E. Solomon and R.A. Flavell, 'Isolation and characterisation of a human collagen α1(I)-like gene from a cosmid library', *Nucleic Acids Res.*, *10* (1982), 1981–94.
37. Meyerowitz, Guild, Prestidge and Hogness, 'A new high-capacity cosmid vector'.
38. B.R. de Saint Vincent, S. Delbrück, W. Eckhart, J. Meinkoth, L. Vitto and G. Wahl, 'The cloning and reintroduction into animal cells of a functional CAD gene, a dominant amplifiable genetic marker', *Cell, 27* (1981), 267–77.
39. T. Lund, F.G. Grosveld and R.A. Flavell, 'Isolation of transforming DNA by cosmid

rescue', *Proc. Natl. Acad. Sci. USA*, *79* (1982), 520–4.

40. Chia, Scott and Rigby, 'The construction of cosmid libraries'.
41. Arrand, Rymo, Walsh, Bjorck, Lindahl and Griffin, 'Molecular cloning'.
42. D.M. Lonsdale, R.D. Thompson and T.P. Hodge, 'The integrated forms of the S1 and S2 DNA elements of maize male sterile mitochondrial DNA are flanked by a large repeated sequence', *Nucleic Acids Res.*, *9* (1981), 3657–69.
43. Chia, Scott and Rigby, 'The construction of cosmid libraries'.
44. P. Baldacci, A. Royal, F. Bréggère, J.P. Abastado, B. Cami, F. Daniel and P. Kourilsky, 'DNA organisation in the chicken lysozyme gene region', *Nucleic Acids Res.*, *9* (1981), 3575–88.
45. Hashimoto-Gotoh, Franklin, Nordheim and Timmis, 'Specific-purpose plasmid cloning vectors'.
46. K.N. Timmis, 'Gene manipulation *in vitro*' in S.W. Glover and D.A. Hopwood *Genetics as a tool in Microbiology*, *Soc. Gen. Microbiol.*, *Symp. 31* (1981) 49–109.
47. T. Hashimoto-Gotoch, F.C.H. Franklin, A. Nordheim and K.N. Timmis, 'Specific-purpose plasmid cloning vectors. I. Low copy number, temperature-sensitive, mobilization-defective pSC101-derived containment vectors', *Gene*, *16* (1981) 227–35.
48. Collins, '*E.coli* plasmids'.
49. Collins and Hohn, 'Cosmids'.
50. Collins and Brüning, 'Plasmids useable as gene-cloning material'.
51. Collins, '*E.coli* plasmids'.
52. Saint Vincent, Delbruck, Eckhart, Meinkoth, Vitto and Wahl, 'The cloning and reintroduction'.
53. Ish-Horowicz and Burke, 'Rapid and efficient cosmid cloning'.
54. Ibid.
55. Hohn and Collins, 'A small cosmid for efficient cloning'.
56. Collins, '*E.coli* plasmids'.
57. Lund, Grosveld and Flavell, 'Isolation of transforming DNA'.
58. D. Hanahan and M. Meselson, 'Plasmid screening at high colony density', *Gene*, *10* (1980), 63–7.
59. Cattaneo, Gorski and Mach, 'Cloning of multiple copies of immunoglobulin genes'.
60. Meyerowitz, Guild, Prestidge and Hogness; 'A new high-capacity cosmid vector'.
61. Schweizer, Case, Dijkstra, Giles and Kushner, 'Identification and characterization of plasmids'.
62. Ish-Horowicz and Burke, 'Rapid and efficient plasmid cloning'.
63. Grosveld, Dahl, Boer and Flavell, 'Isolation of β-globulin-related genes'.
64. Palva and Liljeström, 'Cosmid cloning and transposon mutagenesis'.
65. W. Lindenmaier, H. Hauser, I. Greiser de Wilke and G. Schütz, 'Gene shuttling: moving of cloned DNA into and out of eukaryotic cells', *Nucleic Acids Res.*, *10* (1982), 1243–56.

GLOSSARY

Annealing:	The association of two single-stranded nucleic acid molecules by hydrogen bonding between complementary bases on the respective strands.
Antigen:	A molecule that is capable of stimulating the production of neutralising antibody proteins when injected into a vertebrate.
Antibody:	A particular form of gamma-globulin present in the serum of an animal and produced in response to invasion by an antigen, conferring immunity against subsequent infection by the same antigen.
Autoradiography:	A technique for the detection of radioactively labelled molecules by overlaying the specimen with photographic film. When the film is developed an image is produced which corresponds to the location of the radioactivity.
Complementary DNA:	DNA synthesised by reverse transcription of messenger RNA. The RNA specifies the base sequence of the DNA by the base pairing rules.
Chaotropic:	Resulting in loss of an ordered gel structure, owing to the disruption of hydrogen bonds.
Cistron:	A DNA fragment or portion that specifies or codes for a particular polypeptide.
Clone:	A group of organisms or cells, all originating from a single ancestral organism or cell, and all genetically identical.
Cloning vehicle: (or vector)	Plasmid used for molecular cloning.
Cohesive termini: (Cohesive end)	DNA molecules with single-stranded ends that show complementarity, making it possible, for example, to join end to end with introduced fragments.
Cosmid:	A cloning vehicle (plasmid) consisting of a Col E1 type replicon, joined to the cos λ site of phage lambda DNA. The system combines the advantageous properties of plasmid and phage transduction systems for recombinant work.
Cross-hybridisation:	Hybridisation of a probe to imperfectly matching (less than 100% complementary) molecules.
DNA duplex:	Double-stranded DNA molecule.
DNA polymerase:	Any one of a number of enzymes which catalyse the formation of DNA from deoxyribonucleotides.
Diazotisation:	The activation step leading from the stable unreactive aromatic amine to the reactive unstable aromatic diazo group.
Electro-endosmosis:	The movement of electrophoresis buffer through an ionised support (gel, paper, etc.) towards the electrode of the same charge as the support.
Eukaryote:	An organism or cell with a discrete nucleus.
Exon:	A DNA sequence that is transcribed and codes for final mRNA.
Genome:	The complete set of genes or chromosomes constituting the genetic endowment of an individual.

322

Glycoprotein:	A protein containing a covalently linked carbohydrate moiety.
Heteroduple:	A DNA molecule formed by base pairing between two strands that are not completely complementary.
Hybridisation:	This refers particularly to the formation of a stable double-stranded hybrid molecule by hydrogen bonding between a DNA strand and its complementary RNA. Also used to describe DNA-DNA association.
Hydrophobic:	Literally meaning 'water-hating'. For example, the amino acids valine, isoleucine, leucine all have hydrophobic side chains. When such groups come together their contact with the surrounding aqueous enviroment is reduced and a stable bond results.
Intensifying screen:	A solid layer of a fluorescent compound (e.g. calcium tungstate) which emits visible light when struck by β-particles or γ-rays.
Isocratic elution:	Chromatographic elution with a mobile phase of constant composition.
Intron:	A DNA sequence that is transcribed but does not appear in the mature mRNA transcript.
In vitro:	Literally 'in glass' meaning in the test-tube or outside the organism.
In vivo:	Within the organism.
Kilobase (Kb):	One thousand nucleotides in sequence.
Klenow fragment:	Piece obtained from DNA polymerase 1 by proteolytic cleavage: it lacks the 5' to 3' exonuclease activity.
Lectins:	A group of plant proteins which bind to specific carbohydrates.
Ligation:	Covalent linkage of DNA termini with T4 DNA ligase.
Maxicells:	Heavily irradiated cells with strongly degraded chromosomal DNA. Plasmid-containing maxicells are able to express plasmid-encoded genes.
Minicells:	Small, anucleate, non-growing bodies, almost spherical in shape, produced by aberrant cell division at the polar ends of bacteria. Plasmid-containing minicells are able to express plasmid-encoded genes.
Mismatching:	Regions in a double-stranded nucleic acid molecule where the bases on the respective strands are non-complementary, and therefore do not hydrogen bond.
Mobile phase:	The eluting solvent used in HPLC.
Monoclonal antibody:	Immunoglobulin produced by a single clone of lymphocytes.
Oligo (dT)-cellulose:	Oligo (deoxythymidylic) acid covalently bound to cellulose.
Operon:	A cluster of functionally related genes regulated and transcribed as a unit.
Periplasm:	The fluid in the periplasmic space. The periplasmic space of gram-negative bacteria is the space between the cytoplasmic and outer membrane.
Plasmid:	An extrachromosomal hereditary element consisting, chemically, of a covalently continuous DNA molecule, found in certain bacteria and fungi.
Primer:	A short DNA fragment used for the initiation of DNA synthesis with DNA polymerase.

Poly (U)- *Sepharose*:	Poly (uridylic) acid covalently bound to a proprietary preparation of agarose.
Potter homogeniser:	Precision-made tissue homogeniser comprising a glass tube-shaped mortar and Teflon pestle.
Prokaryote:	A cell or organism which lacks a discrete nucleus.
Probe *(hybridisation)*:	DNA or RNA molecule radiolabelled to a high specific radioactivity, used to detect the presence of a complementary sequence by molecular hybridisation.
Protein A:	Isolated from *Staphylococcus aureus* and binds to the F_c portion of IgG antibodies, without affecting the antigen binding site and as such is an ideal anti-antibody.
Restriction *endonucleases*:	A group of enzymes, commonly found in bacteria, which break internal bonds of DNA at highly specific points.
Reverse transcriptase:	An enzyme which synthesises DNA from an RNA template.
Sedimentation co-efficient:	The velocity of a solute in relation to the centrifugal field strength applied by an ultracentrifuge, expressed in Svedberg units (S) or seconds.
Sequence complexity:	The number of nucleotide residues in a particular nucleic acid sequence.
Spheroplasts:	Bacterial cells with a degraded cell wall and spherical in shape. Spheroplasts are only stable in hypertonic solutions.
Stationary phase:	The immobilised column packing material competing with the mobile phase for the sample molecules migrating through the HPLC column.
Template:	Polynucleotide molecule which acts as a mould for the synthesis of a complementary nucleic acid strand.
Transduction:	Exchange of genetic material mediated by a virus.
Transformation:	The acquisition of new genes or alleles following incorporation of nucleic acid (usually double-stranded DNA) into a cell.
Transposon:	A DNA element which can insert at a very large number of potential sites into plasmids or the bacterial chromosome independently of the host cell recombination system. In addition to genes involved in insertion, transposons carry genes conferring new phenotypes on the host cell, e.g. kanamycin resistance, ampicillin resistance, etc.
Vector *(or cloning vehicle)*:	A self-replicating DNA molecule, usually a small plasmid or bacteriophage, used to introduce a fragment of foreign DNA into a host cell.

NOTES ON CONTRIBUTORS

Stephen A. Boffey: Department of Biological Sciences, The Hatfield Polytechnic PO Box 109, Hatfield, Herts AL10 9AB, England

Jan van Embden: Laboratory of Bacteriology, National Institute of Public Health, PO Box 1, 3720 BA Bilthoven, The Netherlands

Brian G. Forde: Biochemistry Department, Rothamsted Experimental Station, Harpenden, Herts AL5 2JQ, England

Wim Gaastra: Department of Applied Microbial Genetics, Technical University of Denmark, Building 221, 2800 Lyngby, Denmark

Keith Gooderham: MRC Clinical and Population Cytogenetics Unit, Western General Hospital, Crewe Road, Edinburgh EH4 2XU, Scotland

Don Grierson: Department of Physiology and Environmental Science, University of Nottingham School of Agriculture, Sutton Bonington, Loughborough, Leics LE12 5RD, England

Christopher G.P. Mathew: Department of Chemical Pathology, University of Cape Town, Republic of South Africa

Elaine L.V. Mayes: Imperial Cancer Research Fund, Lincolns Inn Fields, London WC2, England

Frits R. Mooi: Department of Microbiology, Biological Laboratory, Free University, de Boelelaan 1087, 1081 HV Amsterdam, The Netherlands

Robert H. Nicolas: Institute of Cancer Research, Royal Cancer Hospital, Chester Beatty Research Institute, Fulham Road, London SW3 6JB, England

Bauke Oudega: Department of Microbiology, Biological Laboratory, Free University, de Boelelaan 1087, 1081 HV Amsterdam, The Netherlands

Robert J. Slater: Department of Biological Sciences, The Hatfield Polytechnic, PO Box 109, Hatfield, Herts AL10 9AB, England

Bryan John Smith: Institute of Cancer Research, Royal Cancer Hospital, Chester Beatty Research Institute, Fulham Road, London SW3 6JB, England

Jim Speirs: CSIRO Division of Food Research and School of Biological Sciences, MacQuarie University, North Ryde, New South Wales 2113, Australia

John M. Walker: Department of Biological Sciences, The Hatfield Polytechnic, PO Box 109, Hatfield, Herts AL10 9AB, England

INDEX